SpringerBriefs in Latin American Studies

More information about this series at http://www.springer.com/series/14332

João Santos Nahum

Modernization and Political Actions in the Brazilian Amazon

The City of Barcarena, Pará

Springer

João Santos Nahum
Instituto de Filosofia e Ciências Humanas
Universidade Federal do Pará
Belém, Pará
Brazil

ISSN 2366-763X ISSN 2366-7648 (electronic)
SpringerBriefs in Latin American Studies
ISBN 978-3-319-58029-6 ISBN 978-3-319-58030-2 (eBook)
DOI 10.1007/978-3-319-58030-2

Library of Congress Control Number: 2017938534

Printed on acid-free paper

This Springer imprint is published by Springer Nature
The registered company is Springer International Publishing AG
The registered company address is: Gewerbestrasse 11, 6330 Cham, Switzerland

Preface

Abstract The central idea of the text is that a system of conservative actions is reproduced in front of the modernization process of the territory of Barcarena. The innovations that the territory of the city knows, either in its territorial configuration or in public administration, are reproduced according to the system of conservative actions, bulwark of the political and economic hegemony of a small decider group. These are deployed according to conservative political practices, reaffirming "the power of delay."

Keywords Modernization · Conservative political actions · Territory used · Amazon

I am happy to present this book, the result of a Ph.D. thesis on the use of the territory in Barcarena, held in May 2006 at the Postgraduate program of Geography at the Institute of Geosciences and Exact Sciences, State University Júlio de Mesquita Filho (UNESP), Campus of Rio Claro, Sao Paulo. The thesis was done under the guidance of Prof. Samira Peduti Kahil, to whom I was the first doctorate advisee and whom Death took so early.

Ten years later is strong the desire to rewrite my thesis, updating and comparing data, interviews, literature, and cartography. I resisted. Allowing myself to only improve the text, clarifying ideas I deem too synthetic to have autonomy of thought and to be self-explanatory. The structuring core of reflection remains intact, *the power of the delay*. For a moment I doubted that I could do the job because the obstacles brought me indignation and paralyzing fears. The conversations with Prof. Samira were always very severe, harsh, and blunt, but there was mutual respect, professional admiration. The last word was hers: *Courage!*

In that tone, texts and research reports and research were built, whose contributions of Prof. Márcio Cataia and Prof. Bernadette Oliveira, in the qualification examination, enriched the reflection so that I could have more clarity about the core of the thesis. Until then, this core was the implementation of the Fiscal Responsibility Law in the territory of Barcarena and the research aimed to show the changes in the management of this territory, which would go through a traditional pattern of planning and public administration to a level of an effective participation of civil society.

The qualification examination, the banking submission concerning showed me that the implementation of this Law is just the tip of the iceberg, *which is the arrival of modernizations in the Amazon region.* These are deployed according to conservative political practices, reaffirming "the power of delay."

After the qualifying examination, I stress the idea that a system of conservative actions is reproduced in front of the modernization process of the territory of Barcarena. The innovations that the territory of the city knows, either in its territorial configuration or in public administration, is reproduced according to the system of conservative actions, bulwark of the political and economic hegemony of a small decider group.

Since then, the task has been to work—boldly—on the fundamentals of the method of analysis and interpretation of geographical space to facilitate, renew my purpose of working the totalizations processes in a part of the national territory. A place: Barcarena. Then, boldly, to choose the variables that best demonstrate my main thesis. All this took me a dose of courage, patience and perseverance. How many versions prior to that I write now! Sometimes I slept in love with the latest version, just to send it to the computer's Recycle Bin the next day. In the process of enfranchisement of this text, myself, geographer, and citizen, when lapidating the text, I was lapidating myself. In a few years, probably, this text and I will not recognize ourselves, therefore, time is king and space its kingdom, but I identify with it now.

Estrangement came after the defense on May 9, 2006. I realized the fool's gold is contained in the thesis. A geographic denunciation having as the main character the uses and abuses of the territory by a traditional management that insisted on upgrading itself to enable more of *the same power of the delay.* I remembered the many reports of people with tanned skin, coarse wrinkles, many toothless and homeless, with nothing beyond the territorial body and sacred temple, so ravished by determined management to make the political bias of fake capital accumulation and, by taking advantage of their naivety, ignorance, and needs of others, reaps goods and expands the patrimony. And we can do nothing! As if we were subjected to the punishment of the vote! As if between the voter and the candidate there was a single and transparent way! As if there were no legal system and production process of political marketing engineering that makes candidates for public office as disposable and recyclable!

At this time we live the Absurd, because the territory used, as affirms Milton Santos, is our picture of life. His understanding is therefore essential to avoid the risk of alienation, the risk of loss of the sense of individual and collective existence, the risk of renunciation of the future. *Indignation is breathed! Because after more than ten years the absurd power of delay reproduces, presiding, supporting and structuring the spatial management and uses of the territory—markedly uneven.* Managers are changed, businesses and enterprises also, but the conservative modernization is the way for the political patronage of Pará to reproduce itself. In my conversations with wistful Samira I said: *Professor, Pará is feudal! Those who can, command, and those who are wise, obey!* She smiled at me. But it is true. This is because politics and space are twin brothers. Every policy has a spatial base and every space has its forms, functions, structures, and processes devised in the field of politics, as well as it is a prerequisite for achieving this.

Understanding the uses of the territory and abuse of it is essential to building a space where citizenship is reinvented, not summarizing the choice to every election of representatives.

In Chap. 1, I start with the reorder of the territory of Barcarena, due to the implementation of the engineering system producer of primary aluminum from Albras/Alunorte. Considering this event (SANTOS 2006), I demarcate two geographical periods in the territorial dynamics of Barcarena. One before and another after the implementation of Albras/Alunorte project. I focus on the second, in which modernization takes the form of territorial reorganization imposed by entrepreneurs and the federal government. As if such modernization project took into account the interests of the entire population of Barcarena!

The reordering actions are propagandized as essential and inevitable progress and local development policies. As if all citizens were benefited! As if this was the only way to modernize the territory of Barcarena!

The territorial reorganization, I will explain in the next chapter, is the foundation of the welfare practice of political coercion on civil society, the centralization of power and the use of the territory as a resort of political domination by the elite city ruler.

In Chap. 2, I reveal that, with the exception of the industrial complex area, the rest of the territory of Barcarena, including the village of Cabanos, is controlled by a system of conservative actions, administrative centralization promoter, lack of transparency and absence of participation of civil society in the preparation of administrative planning. In sum, old production practices of territorial inequality. Also, before upgrading vectors, especially those who come to regulate the processes of decentralization and administrative democratization, the mayor, councilors, and secretaries decide how to implement them in a way that suits their particular interests. Reaffirming that what happens in all of Brazil happens in Barcarena.

In Chap. 3, I examine the reproduction of conservative actions system in each process of modernization of the territory of Barcarena, focusing on the process at the time of implementation of the Fiscal Responsibility Law in management (2001–2004) of Laurival Magno Cunha. This government did not use the law as a fiscal adjustment instrument, as conceived by the global reason. Suited to his interests are the planning, public participation, and transparency established by Law, to show up modern and hide his conservative style, anti-democratic and elitist land management.

During the final considerations I conclude with a summary. The time of completion—and at the same time an insight and foresight—allows me to dispel the belief that it would accomplish much, but at the same time shows me how small is the size of what I have done. Recognizing it motivates me to think about issues that I come across, such as the need to build a map of the power of the elites in the state of Pará, which subsidize the development of geographic analysis to elucidate the influence of traditional political culture in territorial management, even before the administrative modernization imperatives posed by global order.

Belém, Brazil João Santos Nahum

Contents

Chapter 1
The Modern Territorial Reordering of Barcarena

Abstract In this chapter, I start with the reorder of the territory of Barcarena, due to the implementation of the engineering system producer of primary aluminum from Albras/Alunorte. Considering this event (Santos 1999), I demarcate two geographical periods in the territorial dynamics of Barcarena. One before and another after the implementation of Albras/Alunorte project. I focus on the second, in which modernization takes the form of territorial reorganization imposed by entrepreneurs and the federal government. As if such modernization project took into account the interests of the entire population of Barcarena. The reordering actions are propagandized as essential and inevitable progress and local development policies. As if all citizens were benefited. As if this was the only way to modernize the territory of Barcarena. The territorial reorganization, I will explain in the next chapter, is the foundation of the welfare practice of political coercion on civil society, the centralization of power and the use of the territory as a resort of political domination by the elite city ruler.

Keywords Territory · Modernization · Barcarena · Albras/Alunorte

An object stands on the beach of Caripi in Pará River, located east of Itupanema village and at a distance about 5 km from the village of Cabanos—the Tree House. A unique and imposing house built on pillars hidden between two large trees, suggesting the curious glances, that it is suspended in the canopy of these. All of hardwood, it is part of the Hotel Sumauma, set of B&Bs whose style is reminiscent of the dwellings made to withstand snow and cold weather. The hotel makes up the landscape architected from the primary aluminum producer venture called Albras/Alunorte. According to these companies, the territorial configuration is reorganized, the social dynamics as well, in sum, the territory used or the space geographical of Barcarena.

Before installation of these companies, Caripi was a place that had no infrastructure except for the small boats that made the river transport to this beach, so it was uncrowded. In the late 1970s and early 1980s, to the extent that was added to the territorial configuration of Barcarena: ports, roads, planned urban center,

J.S. Nahum, *Modernization and Political Actions in the Brazilian Amazon*,
SpringerBriefs in Latin American Studies, DOI 10.1007/978-3-319-58030-2_1

factory, power system, the beach can be seen transformed into the village recreation area of Cabanos. In Caripi the following are installed: a power grid, running water, telephone, hotel services, bars, restaurants, security, lifeguards, cleaning, and garbage collection. This differentiates the place from every other in the Barcarena coastline, making it part of the tourist routes. The frequency of visitors increased from 2002, with the connection of Belém to the PA-150 highway and the village of Conde port, through the Road strap, including Caripi in the supply chain of Pará.

Leaving Caripi, moving toward the southwest, after passing through the Itupanema village, we find the village of Conde. Quiet place and with a spatial configuration similar to many villages of the Amazon: a church, simple wooden houses, or masonry in different models. Some eternally unfinished and many skirting the Square of São João Batista. In the village, near the beach, stood out the stigmatized place of sex workers, with its dark streets, gravel and narrow, where some bars remain—in most cases they boil down to simple houses where, in the front room there is one desk, shelves with drinks and cigarettes, a pool table with poor lightning and an owner who greets the regulars. These shops, usually, without any refinement in terms of menu, service, health, and safety, preceded the entry hall in which work the agency of prostitutes. There were, however, refined places where the quality of services was proportional to the customer's purchasing power, especially if he was a foreigner. In that case, it was possible for the manager of the show house to put up for auction some "night girl" (Dimenstein 1995). In the late 1970s and early 1980s, the area surrounding the beach in Vila do Conde made itself unique as a leisure zone, where the *fichado*, the *pawn*—local migrant codenames—spent part of their money.

Before installing the complex of Albras/Alunorte, Vila do Conde was a place where the economic activities were hunting, fishing, gathering fruits, and planting cassava. As in other places in Barcarena, residents knew each other's origin, the social relations celebrating bonds of solidarity. The productive activity had its foundations, predominantly, in family sites, usually small farms sustained in social relations of self-employed production. Those who did not have land to produce, were engaged in carpentry, construction, small shops, and general services. Currently, in Vila do Conde few families practice the small agricultural production, the predominant work is in the urban area in the core village of Cabanos, the seat of the municipality or in another location.

Another part of the Barcarena territory is the town of São Francisco, on the right bank of the river Barcarena. In this place, the Jesuits in the eighteenth century built the church of São Francisco, the first of the municipality. In 1943, when Barcarena was emancipated from Belém, the first head office was the town of São Francisco. Due to the small navigation flow in the Barcarena river, the place faced economic difficulties. The elevation to city status was accompanied by the aspiration for territorial change. Traders and the general population complained for being displaced from the economic circulation axis leading to Belém.

Starting in 1946, the mayor Frederico Duarte de Vasconcelos made public the idea of changing the municipal seat to the left bank of the River Mucuruça. Despite the stubborn conviction about the location for the headquarters transference, political discussions were very intense. This process lasted from 1946 until 1952, often receiving broad support from local opposition groups, others included in the political platform of parties in election periods to be approved by State Decree by the Federal Interventor in 1952. According to Allowances For A Study of History of the Municipality of Barcarena (Subsídios Para Um Estudo da História do Município de Barcarena), the dream of the residents and local politicians took legal form through the Municipal Law No. 71 of 10/29/52 and the State of Law No. 534 of 23/08/53. Only with the shift in the center of the municipal administration and the building of the town hall in the second management of Raimundo Alves Dias (1959–1963) the headquarters transference project acquired spatiality.

In the first administration of Mayor Laurival Campos Cunha (1963–1967), the territorial configuration of the new district headquarters acquired contour. The government granted land and building materials to residents of the village of São Francisco, popularly called Old Barcarena, encouraged them to move to the new district headquarters where the buildings of the city and the legislative chamber, the forum, the station, the bus station, and the shopping area were erected. In short, a set of geographic objects shaped the place as the center of political and economic life. The function of the village of São Francisco as the municipal administrative headquarters ended.

Today, still quiet, the everyday life of São Francisco only changes its feature and stirs in the times of círio and arraial, names of commemorative religious festivities of their patron saint. The place is marked by simple houses of wood or masonry, not all featuring running water or electricity, much less telephone and public transport. The services public are in deficit, sewage treatment is almost nonexistent and quality of gravel streets covered with tar and potholes indicate the precariousness of public services. Deficiency in the health and public safety services also adds to the list.

The village of São Francisco, village of Conde, and Caripi are places (Map 1.1) comprising systems of objects that "have autonomy of existence for the things that form them: streets, buildings, plumbing, industries, businesses, restaurants, electric parts, cobbles, but have no autonomy in the stock systems because every day new shares replace the old, they are imposed and exercised" (Santos 1991, p. 52, author's translation). Such places are fractions of the territory of Barcarena, and therefore have no significance of autonomy since the nature of this space—be it the dimension of landscape, territorial configuration, or social dynamics—is related to what happens elsewhere in Barcarena, Pará, Brazil and in the world. Places which are witnesses of social morphology, territorial inequalities produced from the event (Santos 1999) conducted by a large enterprise, the Albras/Alunorte.

The whole process of territorial reorganization required for the implementation of these companies—such as the construction of production systems, energy, transportation and communication, a planned urban center, the spontaneous occupation of places and growth of municipal revenues—are part of a fabulous modernization made of simulations and devices more concerned with modern look than to be modern, the result of a system with conservative political action using the territory as a resource to ensure the interests of the small circle of agents who decide

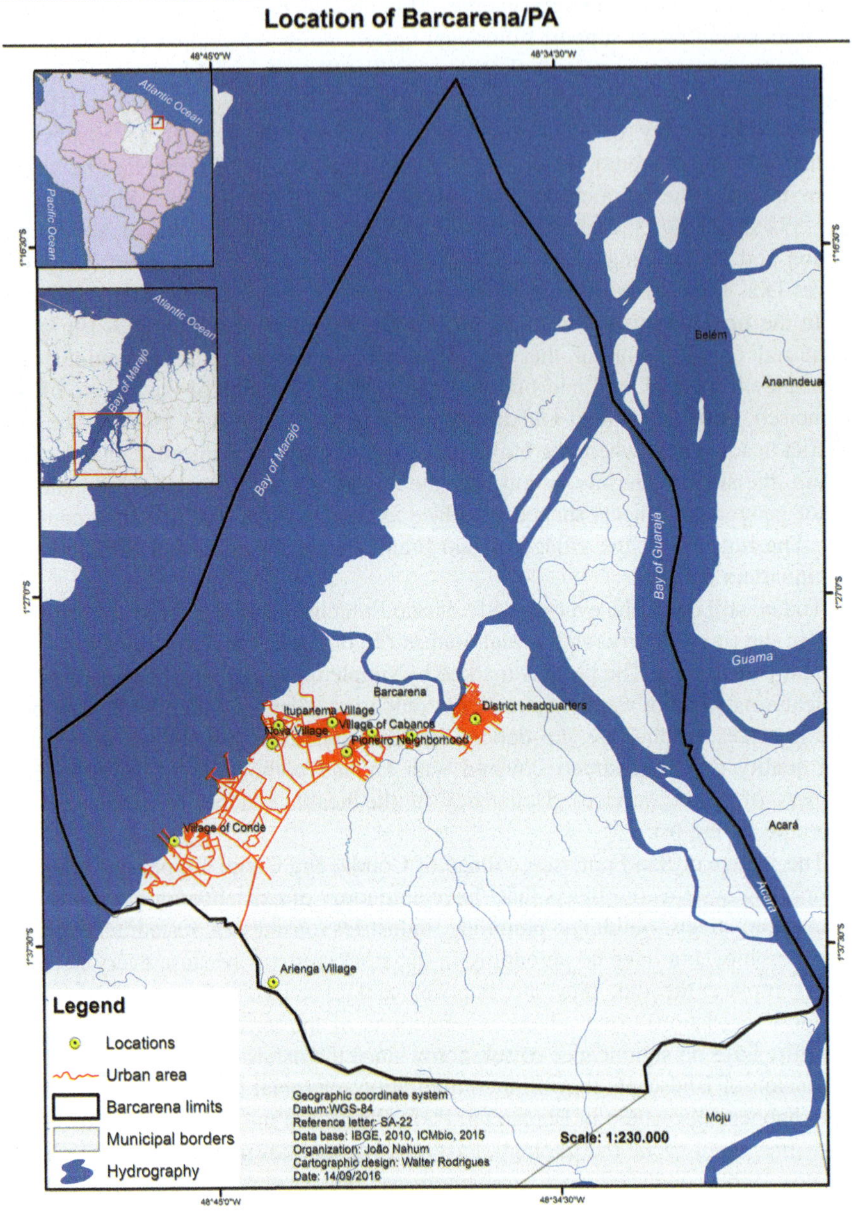

Map 1.1 Spatial configuration and areas of spontaneous occupation of Barcarena-PA

(businesses and elite municipal governing), as conceptualized by Santos (1999) "by those who can choose what will be widespread and more, those able to choose the action that, in this sense, will hold" (p. 65, author's plural and translation).

The reordering actions are propagandized as essential and inevitable progress and development policies. As if all citizens were to benefit. As if this was the only way to modernize the territory of Barcarena. But it contributes to the reproduction of privileges and interests of the economic and political elite. The territorial reorganization—we explain below—is the spatial foundation of welfare practice of political coercion on civil society, the centralization of power and the use of the territory as a resort of domination by the elite city ruler.

1.1 The Modernization Through Albras/Alunorte

In the 1970s, the increase in oil prices, followed by its cartelized operation through the Organization of Exporting Countries and Petroleum Exporting Countries (OPEC), restructured the division of labor globally. For Santos (1999, p. 106) the division of labor is "a process by which resources are distributed socially and geographically." The spatial reorganization triggered with the transfer of industrial plants producing primary aluminum, managed by companies and Japanese capital conglomerates to territories that offered better competitive advantages.

Among the comparative advantages for the establishment of these foreign objects, counted in favor of Pará territory, is the presence of bauxite deposits in the Trombetas River, in the municipality of Oriximiná, known since 1963. But the real value of these resources does not "depend on its separate existence, but on its geographical qualification, i.e. the joint meaning that each and every one obtain by the fact of participating in one place" (Santos 1999, p. 107, author's translation). The geographical qualification of these mineral resources are built from the invitation that the Brazilian government made to the Japanese group of *Aluminium Resouces Development Co* (ARDECO) to collaborate in carrying out studies for the production of alumina in the Amazon region with the Vale do Rio Doce Company (CVRD), back then a state-owned company not fully privatized.

The use of natural potential as an economic resource has intensified in the post-1964 regime, culminating in "the re-creation, by the military, of the Amazon as a frontier" (Ianni 1981, p. 141, author's translation), by the Amazon Operation in 1966–1967. This operation craved to integrate in the Brazilian Amazon the territorial division of labor. The military government created the Superintendent of the Amazon Development—SUDAM (Law 5,173, of October 27, 1966, as amended and innovated by Law 5,373 of December 7, 1967); Banco da Amazônia S/A—BASA (Law 5122 of September 28, 1966); the Manaus Free Trade Zone (6.1244 Law of 28 August 1967, regulated by Decree-Law 288 of 28 February 1967); and legislation on tax incentives (Law 5.174, of October 27, 1966).

The Brazilian state, sustained in a conservative action system geographically described the Amazon resources through legislative, financial and technical conditions to extend to this part of the country its modernization model that combined "geopolitical [interests] with extensive development of capitalism in the region" (Ianni 1981, p. 141, author's translation). In reality, this model caused the "military governments to deeply undermine the state apparatus with the interests of foreign and domestic monopoly of capital" (Ianni 1986, p. 16, author's translation). An example of this commitment was the privatization, denationalization, inside the Company Vale do Rio Doce, as a result of intense association of "state" with multinational groups, especially in the decades of 1970–1980. According to Oliveira (1993, pp. 48–49, author's translation), "through 17 companies, CVRD is affiliated with other national and international groups [...] constituting a kind of articulation of state/multinational."

The state assumed the role of guarantor of the internalization of external interests, represented in the Amazon, among others, by large projects and mining-metallurgical enterprises and hydroelectric. Through a system of conservative actions, the state acted as an entrepreneur, directed the public fund to create territorial conditions favorable to investments in projects related to forestry, fisheries, agriculture, the wood processing, but above all the major works by energy infrastructure and mining-metallurgical production. In Brazil, paradoxically, the public fund "has to be the assumption of capital accumulation of financing on one side and the other, the financing of the reproduction of the labor force" (Oliveira 1998, pp. 19–20, author's translation).

The federal government offered to entrepreneurs Albras/Alunorte a number of incentives to attract to the region the industrial plant of aluminum. Starting with the energy, not only by the amount required for aluminum production. In September 1973 is created the Eletronorte—Power Plants in North Brazil S/A, in order to expand the exploitation of the energy potential of the region and build a large engineering system, the hydroelectric plant of Tucuruí energy supplier for ore and metallurgical production of major projects (Table 1.1).

> In fact, when the construction of the dam was decided in 1974, it was only the appendix of one of the largest industrial complexes in the world aluminum, Albras/Alunorte [...] Alone, this enterprise would require no less than 1.3 million KW, seven times more than Belém, with its first million inhabitants currently consumes.
>
> But when the Japanese were to calculate the cost of the dam, which would be 300 kilometers away from the two factories were frightened and asked the Brazilian government to ease the consortium of such enormous weight. The plant, demanding $2.5 billion, would cost more than the industrial plants themselves. The government transferred then the burden for Eletrobras, who decided to run it using international loans.

Table 1.1 Main raw materials

Alumina (Al_2O_3)	1920 kg
Coke (C)	400 kg
Tar (C)	100 kg
Aluminum Fluoride (AlF_3)	20 kg
Electricity (CC)	14,000 kW h

Source Author

Table 1.2 Resource application of the support program for industrial complex of Barcarena-para. Position In: 1987: US $1000.00

Subprogram	Value	Part (%)
Projects and roadworks	3552.7	4.0
Projects and port and waterways works	30,369.0	34.0
Projects and implementation of works in the urban center at village of Cabanos	36,021,3	40.3
Projects and works for the catchment area of the industrial complex	4970.4	5.5
Others	4172.7	4.6
Management and monitoring	6769.0	7.6
Operation and maintenance	3383.5	3.8
Technical consultancy and supervision	1648	0.2
Grand total	89,403.4	100.0

Source Lôbo (1996, p. 127, author's translation)

> During this period, however, the dimensions of the Albras/Alunorte were halved and it "only" require 644 KW, still 5% of all energy consumption in Brazil, which assures the title of largest individual consumer of the country's energy [...] (Pinto 1982, p. 19, author's translation).

The Brazilian state—as well as having financed the Tucuruí Hydroelectric Plant placing it at the disposal of Albras/Alunorte—to facilitate the establishment of the great project, funded and built transportation of systems engineering in the waterway and road sector, a grain port and general cargo for large draft ships, in addition to the urban core (Table 1.2).

For Lôbo (1996), the range of benefits provided by the state to aluminum entrepreneurs was extended when the project Albras/Alunorte, which was born in the Amazon III Development Plan 1980–1985, incorporated in 1981 all the enterprises of the Great's Carajás Program (PGC) and received special treatment by the organs and entities of the Federal Administration for the purpose of:

1. Concession, lease and titling of public lands, regulation and discrimination of vacant land or, when appropriate, expropriation of land necessary for the implementation of the projects;
2. License or concession for construction and operation of port facilities;
3. Contracts for the supply of electricity and water transport;
4. Transfer or lease of rights of mineral or forestry exploration;
5. Authorization, issuance of guidelines and grant to finance exports;
6. Authorization and issuance of guidelines for import of machinery, equipment and instruments, as well as assemblies, parts, accessories, for the deployment, expansion, modernization or re-equipment of enterprises, including in the case of foreign direct investment, in the form of goods or services;
7. Authorization and registration of foreign loans, including for the payment abroad of goods or services;

8. Approval of grant or guarantee of the National Treasury or public financial institutions for loans abroad;
9. Authorization for financing of mining companies;
10. Participation, with public funds, in the capital of the respective companies of the projects;
11. Approval of technology transfer agreements, assistance or technical advice for the implementation and operation of projects (Lôbo 1996, p. 139).

The conservative actions system created conditions for ALBRAS—Brazilian Aluminum S/A—to settle in Barcarena. The company that produces primary aluminum currently has the largest installed capacity in Brazil. Albras was implemented in the late 1970s, the result of a bilateral agreement between the governments of Brazil and Japan. It has as shareholders: Company Vale do Rio Doce (CVRD), with 51% of the capital represented by their subsidiary Vale Rio Doce Aluminum S/A (ALUVALE)—at that time CVRD was not fully privatized, which would occur in 1997—and Nippon Amazon Aluminium Co. Ltd. (NAAC), an association of 32 Japanese companies, with 49% of the capital. The total investment for the implementation of Albras was $1.5 billion.

The area defined for the installation of the industrial complex aluminum producer in Barcarena, located 7 km from the county seat, occupies 40.000 ha, distributed as follows: (a) *the industrial area*: includes the port located in Ponta Grossa, in Vila do Conde; area of Alunorte, near the port; the area of Albras, located on the harbor front and the continuous batch of Alunorte; the area of Eletronorte, located next to the lot of Albras, for installation of the substation which lowers the voltage coming from the Tucuruí; and the expansion area, made available to the Industrial Districts Company of Pará—CDI, for the installation of the Industrial District; (b) *the urban expansion area:* reserved to meet the demand for industrial lots; (c) *the transition area*: intended to protect the Ecological Reserve and support the diverse activities of an industrial character; (d) *the deployment area of the New Urban Center of Barcarena*: where the homes of workers of Albras/Alunorte are located.

Then the production system was implanted in two phases, each with a nominal capacity of 160 thousand tons/year. The first began operating in July 1985 and the second reached full capacity in early 1991. Throughout time, after the implementation of successive technological improvements, the plant reached the capacity of 360 thousand tons/year and with the expansion held in 2001, the annual output reached 406,000 tons of metal.

The choice of Albras/Alunorte complex area took into account: (1) the relative proximity of the sources of its main raw materials (bauxite and electricity); (2) the possibility of building and use of a port to allow the berthing of large-capacity vessels above 40,000 tons; (3) soil features with flat topography and favorable basement to support the plant foundations; (4) the existence of water of high quality and volume for industrial use; (5) the cheap labor availability; (6) low density area occupation and at the same time, near a large urban center. After analyzing the factors and given the alternatives, it was decided that it should stay in the area of influence of Belém. Barcarena was the territory which brought together many of these advantageous features.

In one part of the Barcarena territory was installed a technical object system, associated with technical actions systems, which are jointly and inseparably instruments of relationships and similarities between places and regions, be it in the local scale or between countries. In other terms, such as production today is fragmented within the company and between companies, which requires "a joint political command and this command policy unit functions within firms" (Santos 2000, pp. 26–27, author's translation) also there is the fragmentation of the territory, hence the need for centralization, a remote control for the now scattered actions. Let us see:

(a) Electricity[1]:

The electricity used is generated by Eletronorte, in the Tucuruí Hydroelectric Plant, located on the Tocantins River in the municipality of Tucuruí, 300 km from the plant, and is provided for Albras in a subsidized way.[2] It is responsible for the use of more than 15% of the generation of this plant, whose current capacity is 8340 MW. The energy reaches the voltage of 500 kV in the substation of Vila do Conde, which is lowered to the 230 kV that is supplied to the company.

(b) Raw materials:

1. Alumina: basic raw material for aluminum industrial production is obtained from the processing of bauxite ore, abundant in the Trombetas River region in northwestern Para. In the first ten years of operation, Albras used imported alumina from various countries like the United States, Suriname and Venezuela. However, with the start of production of North Alumina do Brazil S/A (ALUNORTE),[3] in 1995, this raw material is provided directly by the company. The bauxite produced by Mineração Rio do Norte is carried for 1000 km along the rivers Trombetas and Amazon and arrives at the port of Vila do Conde, where it is conducted to Alunorte for refining; it being extracted alumina fueling Albras and exports to other aluminum industries in Brazil and abroad.
2. Aluminum fluoride is a product acquired by Albras in the international market, mainly large manufacturers in Mexico and Tunisia. Fluoride has the appearance of a whitish powder and is received in sacks or big bags at Vila do Conde port.

[1]Information about the components of this system, especially concerning the production process, are available at: www.albras.net. Access: 07 June 2004.

[2]Lúcio Flávio Pinto in *Albras and Eletronorte signed the country's largest energy contract*, cites some terms of the new contract between Eletronorte and Albras for the power supply. The joint venture Albras will pay R $53.00 per MWh. View Source: Personal Journal, 21/05/2004. Available at: www.amazonia.org.br/noticia/noticia. Access: 07 June 2004.

[3]Alunorte has in its ownership structure: CVRD—Company Vale do Rio Doce S.A: 57.03%; Norsk ASA: 34.03%; NAAC—Nippon Amazon Aluminum Co.: 3.80%; CBA—Brazilian Aluminum: 3.6%; JAIC—Japan Alunort Investment Co.: 19%; Mitsui & Co.: 0.23% and Mitsubischi Corporation: 0.10%. Source: www.alunorte.net. Access: 07 June 2004.

3. Petroleum coke: main component for the manufacture of anode coke, it is obtained by calcining the green coke originated from oil refineries; it has the aspect of granulated charcoal; it is received in bulk from major suppliers—mainly the US and India.
4. The electrolytic tar: obtained from the distillation of coal tar, is used in the manufacture of anodes as a binder of the particles of coke. It is provided in solid form—called pencil tar—basically from European, German, and French companies.
5. Heavy oil, BPF, or Type A: Albras uses large quantities of oil in baking ovens, it serves as fuel for combustion and heat generation required to produce baked anodes with special properties required by the electrolytic process, among them high mechanical strength and low electrical resistivity. The oil consumed by Albras is provided directly by two major oil companies to close the factory premises in the Vila do Conde port.

(c) Production technology:

Albras adopted (for its 960 ovens) a technology of Mitsui Aluminium Co. Ltd., Japan, which uses pre-baked anodes produced in the industrial plant of the company.

The key inputs required to produce a ton of primary aluminum are (Table 1.1).

The final product of Albras is a 22 kg aluminum ingot, with minimum purity of 99.7% Al. The Albras smelter has equipment that enables the production of aluminum alloys and special ingots up to 500 kg. The production meets local industries, but mostly is for the foreign market. Aluminum is exported through the port of Vila do Conde, supplying the markets of Japan, Europe, and the United States.

(d) Transport systems:

The Vila do Conde port was designed to serve the alumina-aluminum production complex and the industrial district planned for the area, which can accommodate ships of up to 75,000 tons. It is 1 km from the Albras plant. The Vila do Conde port is a public harbor owned by the Company Docas do Pará (CDP) and in 2002 was part of the road network of the state. In addition to serving the plant Albras, which exports 97% of its production and imports almost all raw materials, the port is adapted to unload bauxite, caustic soda, and fuel oil, also used for loading ships with hydrate and alumina and other types of cargo, including serving the domestic market.

The path from Belém to Albras/Alunorte and to Vila dos Cabanos is through inland waterways, road and river and route. In the carriage of passengers and small loads, the most practical alternatives are speedboats (50 min) or small boats (2 h), leaving Bethlehem two different ports: Arapari Navigation and Navigation Foca. The destinations are the ports of São Francisco and Cafezal, near Vila dos Cabanos and Albras/Alunorte.

Another option is the road and inland waterway transport carried out by a single company, Arapari Navigation. There are five daily trips that begin by crossing the

Bay of Guajará from Bethlehem to the island of Arapari in large boats, taking approximately 50 min. From Arapari to the port to the Albras and Vila dos Cabanos, the journey continues on buses, for about 40 min. For cargo and passenger vehicles, access can be via ferry boats or ferries that leave Bethlehem every hour, and the crossing to the island of Arapari lasts an hour. Then the road to Albras and Vila dos Cabanos, at distance about 40 km, can be taken. More recently, in September 2002 the Road strap was opened that starts on the BR-316 highway—near Belém—allowing access entirely by road to Albras and Vila dos Cabanos.

(e) The urban center of Vila dos Cabanos:

A residential center was created to serve the employees of Albras and other companies that have settled in the industrial area. The urban design and implementation of the village were given to the Barcarena Development Company (codebar), linked to the federal government, which held the ownership of the land, marketed urban land and watched over by obedience to the development plan of the town. Then the Vila dos Cabanos was designed as an "open city," equipped with all basic infrastructure: water, electricity, telephone, and sewage.

The industrial plant, the energy and raw materials supply systems, the transport network, the port, the urban core and expansion areas, in short, the whole system of technical objects we can as Santos (2005, p. 134) call "a point or spot technical-scientific-informational environment." This part of the territorial configuration of Barcarena, markedly modern content is associated with the world of time, "which is the time of multinational companies and supranational institutions" (Santos 1999, p. 111). While for most of the population remain the same old problems given the uneven use of the territory, on the one hand, part of the territory modernized as a resource for the wealth of a few. Throughout the territory, most suffer from lack of social investments. All this contributes to the fragmentation of the territory while deepening the divide between social classes, which requires greater control on the ineluctable social pressure, by the elite.

In the territory of Barcarena, the population living in places for the construction of the industrial complex still suffers the consequences of the conservative political action system. The state and corporate power understand this population as an obstacle to the modernization process and its removal as the required price. The population sought to resist removal processes. But resistance has become fragile in the face of an entire conviction effort undertaken by both state power and by corporate power, spreading the fabulous benefits the project would bring to Barcarena.

During process of removing the population, the state action has taken all normative appeal of the Institute of Lands of Pará (ITERPA), the Institute of Colonization and Agrarian Reform (INCRA), Industrial Development Company of Pará (CDI), and the Company Barcarena Development (CODEBAR).

> To be implemented the legal mechanism of expropriations was designated a state institution, CDI, which held 404 expropriations in the period of 1983-1984 in an area of 40,000 ha, and federal, codebar, who made 155 expropriations in an area of 60,104 ha in

> the period from 1983 to 1984. The area expropriated by the CDI allocated to the implementation of the industrial complex and the construction of the port and the codebar the implementation of Barcarena Center. The expropriations were conducted at different times, at different paces and with varying criteria. Of course this apparent disorganization aimed to hinder a possible articulation of the population (Maia and Moura 1995, p. 246, author's translation).

Following the perspective of prevailing the gap in the first three Amazonian development plans (Nahum 1999), the government of the state of Pará declared (by Decree No. 10064 of April 25, 1977) public utility for expropriation purposes, buildings, and improvements located in Barcarena, in order to implement the project Albras/Alunorte.

Initially, according to Maia and Moura (1995), the Society of Para Industrial Development (CDI) and then the Barcarena Development Company (CODEBAR) —disregarding the familiar place as a production unit and social relations specific to them—stipulated for each of the families, small urban lots of 9 × 30 m or 15 × 30 m, or rural lots of only 5 ha. Unchained up an entire rural and urban restructuring that transformed the colonist producer into a seated consumer.

The agricultural colony Bacuri was a settlement devised by technicians CODEBAR and shows how the planning of this place denounces the distance between idealization of those who plan and the reality of those affected living in the territory for planning. The structure and division of land followed the model of rural lots with houses. Settlers would receive guidance provided by experts appointed by Albras, to produce horticultural intended to supply the urban center of Vila dos Cabanos. This objective miscarried because the relocated had no experience in these activities. They resisted producing other crops, which led many to sell lots. Those who stayed were subsequently able to produce what they wanted and were familiar with.

The dissatisfaction of households with the amount received in compensation has taken collectively the Association of expropriated Barcarena (ADEBAR), founded in 1986. The Association, according to Vasconcelos (1996), formed by 494 families evicted from deployment areas industrial complex, aimed to gather all the expropriated and pass the expropriation experiences for other threatened residents of new expropriations sites in the city. But the primary objective was to claim the monetary losses from the undervaluation of properties and improvements to the CDI and the CODEBAR. According to A. Vasconcelos (1996), the 494 families affiliated to ADEBAR, is addressed to the following locations (Table 1.3).

Regardless of achieving its objectives, ADEBAR was the organized voice of resistance to new expropriations led by state and corporate actions. The association brought together the dissatisfactions that were individual, isolated actions like those seen in the Agricultural Colony of Bacuri, where the settlers refused to carry on the crops and grange activities as designed by the company's technicians, the point of this have to change plans.

Following the example of Adebar came the Community of Integral Agricultural of Barcarena (CAICB), based in Arienga, Km 21 of the PA 151. The highway center brought together 180 families around a project involving the parish and the

Table 1.3 Destination of expropriated families according to the number of families in each location

Locality	Number of families	%
Laranjal	65	13.2
Vila Nova	74	15.0
Agricultural colony of Bacuri	80	16.2
Barcarena	123	24.9
Vila do Conde	65	13.1
Belém	27	05.5
Others	60	12.1
TOTAL	494	100

Source Association of expropriated in Barcarena—ADEBAR apud Vasconcelos (1996, p. 30)

Rural Union of Barcarena. According to Teixeira and Oliveira (1995), the project was part of the social program of Abaetetuba Diocese and was directed at the proposal of the Basic Ecclesial Communities (BECs); Italian financing and has set up an experimental field of agricultural education with the participation of families in 19 coastal villages. The prospect was secure them in the land through agricultural education and training of an organized resistance to the new expropriations that were taking place in Arienga and near the urban core.

The need for organization and resistance occurred because the implementation of the engineering system of Albras/Alunorte fragmented local production unit, the sites. These comprised small property sizes on average 20 ha, with houses built by and for families. The houses were "predominantly of wood, some of clay, covered with straw and/or tile; most had 02 compartments with an area ranging from 30 to 90 m^2. The toilet was outside the house, usually in open air" (Maia and Moura 1995, p. 234). On the ground, the peasants, born and raised in place, caring for fruit trees, developed the rudimentary farming of cassava, and function of this product, raised the flour mill and dug the water well. The use of both the rivers and the forests was collective and had not surrounded that delineate therefore solidarity networks allowed the settlers know where starting a site and ended another. From cassava, besides the manufacture of flours, tucupi and tapioca were obtained, all integrating their diet. To complement the family income they collected fruits, hunted wild animals, fished, and raised chickens and pigs.

In this natural environment modern technology was unusual. And if there was any, it was confined to electricity. Hence the occupation of the whole family. Men, women, elderly, adults, youth, and children employed their days in getting products for their livelihood. The unity between land life and land work remained, which acquired a deep sense of socializing; through it the youngest grew and were educated. The sociability of work organized task forces in which families were associated to perform, make scuffed, making coal, build houses, bridges and piers and clear streams. According to Ponte (1985 apud Maia and Moura 1995, p. 242), "the task force process predominated fellowship as a form of sociability. There was no form of payment that characterized the sale of labor power." The relationship of mutual aid guaranteed to its rightful owner the right of the sale of product and was sure to count on the solidarity of relatives and friends when needed. The task force was a means of mestizo peasant-riverine develop for other dimensions of social life. Where the task force work was performed friendships were strengthened, relationships created and information exchanged.

In farms, the social nature of work was paced by natural time, a different temporality from that prevailing in the technical systems of the Albras/Alunorte, revealing "that each division of labor creates a time its own, different from the previous [...] This is how from each agent, each class or social group, to establish the temporality [...] that is the matrix of the spatiality lived in each place" (Santos 1999, pp. 109–110). The coastal farmer used the territory as a large pantry, where he planted cassava, collected fruits, herbs, roots, seeds, shrimp is fished, the fish scales, and fish skin, all in accordance with the cycles of life of the mineral, plant, and animal; the prevailing weather was wild and not the scientific-informational-technical brought by multinational companies and supranational institutions. However, even if the work had a social character, its fruits were divided among families; and the movement of production in the commercial circuit allowed the reproduction of production conditions. The money was not the main mediator and regulator of social relations, the value of universal exchange enabled families to purchase what they did not produce or did not collect, such as kerosene, diesel, clothes, batteries, and medicines.

In coastal places, away from the urban area of Barcarena and with poor transport system, people get goods through the *marreteiros* of regatões. Kind of traveling salesman of the Amazon rivers, the *marreteiro* is a middleman trader; through him the riverside family unit sells surplus production, enabling it to reach Belém. His vessels, river traders, took/take the grocery function where he was/is, be it meat, salted fish, hammocks, sugar, rum, tobacco, biscuit, bread, clothing, fabrics, shoes, tools, among other things, and the customer could/can still make his or her orders. These merchants have spread through the dense river network of riverside Amazon vivifying the commercial capital circuit, buying in Belém and Manaus at low prizes and reselling in the localities of the riverine to high price, profiting on capital employed initially in the form of money.

In summary, the spatial dynamics of places like Vila do Conde, Vila São Francisco, the Itupanema village, Caripi village, in short, many places of barcarenense territory—before the arrival of Albras/Alunorte—was marked by the profound unity between the family activity and economic activity. The unit's break from the implementation of modern primary aluminum producer object system was one of the spatial consequences for the farmer. This fragmentation of land use change concerns and relationships in everyday life of the former farmer and job search has replaced the family and community work. Rather, according to Maria and Moura (1995, p. 237), "the money was a supplement that served to purchase certain products, which they did not produce [...] Today, without access to land, transformed into wage, they are forced to buy the products necessary to maintain of the family."

The peasant dispossession means deculturation (Santos 1999), as the expropriation of their land destroyed the family unit of production and societal ties were broken. The various stages, from land preparation, planting, maintenance, harvesting, through the technical division of the cassava flour production work, with its derivatives, all fed one intense socialization process, where the flour mill was the aggregator center. These processes supported and were supported by extended ties of solidarity beyond the relationships among families, neighbors embraced, forming communities and were present in negotiations with traders.

At the ranch ruled the idea that the land, rivers, springs, work, everything was used as the common good. For families who were expropriated, the memory of the place where they lived became the memory of community life organized under the family bases. In the new urban community centers are not structured on the old bonds of solidarity, they have become fertile ground for breeding of traditional political culture of Barcarenenses elites, understood as "a set of attitudes, norms, beliefs, more or less widely shared by members of a particular social unit and having as object, political phenomena" (Bobbio et al. 1991, pp. 306–308, author's translation).

The ranches were eliminated for this fraction of the territory to establish the complex of Albras/Alunorte, driving the industrial district of Barcarena, as it was conceived during the military dictatorship in the Amazon Development Plans. At this time, state and federal governments guaranteed the spatiality of exogenous interests to place, behaved like deciding agents because they chose what "will be widespread and more, the action in this sense is to take place" (Santos 1999, p. 65). The state, in conducting the expropriation procedures, supported the action of the law as opposed to the land usufruct of legitimacy by settlers, requiring them the scriptures of the properties. Therefore, in expropriation procedures and compensation was undervaluation of family units, because, in the eyes of the state and corporate power peasants were squatters and not legal owners. In the words of a director of the Rural Workers Union of Barcarena:

> […] here with the implementation of Albras/Alunorte 1700 families were expropriated, being paid miserable indemnifications that even today they fight in federal court in Brasilia for reparations from 1980 and 1985 which they still have not received. Certain personnel, having died, left their children and descendants. But some are still old men and are still hoping to receive compensation which they have not had. This in the deployment of Albras/Alunorte. Since then in the implementation of the other companies the same thing happens, the company buys a piece of land or CODEBAR or CDI, which are the companies that manage public lands expropriated by the federal government and the state government, they are the owners of these areas. And within these areas there are old settlers, who were born and raised there, and every time that a company comes and claims these areas, they lose everything they have due to the company's arrival. And the indemnities are miserable, these people only thicken the city, increase poverty and hunger, because there are no jobs, those who work in businesses are only those who have jobs and they are mostly people coming from outside, because here in the Para region there are few who get a job, a work, here in the Barcarena municipality is just a little more yet (Verbal information).[4]

The municipal administration of José Pinheiro Rodrigues (1977–1983) only helped the reorganization of the landscape, the spatial configuration, and the social

[4]An interview given in March 2005.

dynamics generated by this global command event. Given the hegemonic power of venture capital and standardization imposed by federal and state governments, we obtained the complicity of anxious municipal management by increasing the amount of constitutional transfers resulting therefrom (Table 1.4).

In the territory of Barcarena is reproduced the conservative modernization, according to Becker and Egler (1994, p. 33), "Latin American path to modernity, where the state negotiates with privileged groups to maintain privileges and their inclusion or exclusion in the appropriation of public assets, in exchange for up modernization support down." The modernization of the territory via the implementation of complex Albras/Alunorte was presented as the spatial dynamics of Barcarena depended on the territorial reorganization required to achieve the design of these companies. In order to appear to respect the history and social memory, the names of the leaders of the revolutionary movement Cabanagem were appropriate to baptize the streets, squares, schools in the urban core, symbolically called village of Cabanos. An artifice, a fabulous integration as the modernization proposal by the political elite and its agreements with the companies were proposals and answers to the expectations and needs of the whole society Barcarenense. According to the councilor Manoel Furtado,

> We did not have effective participation in discussions on the implementation of Albras and Alunorte. What they did to us was to take us to Rio de Janeiro so we knew the project. They explained everything to us and told us we had to approve everything, it would be good for Barcarena. Even some things we - the city council – we voted without discussion (Verbal information).[5]

The fabulous speech about increasing the number of direct jobs and growth of the municipal economy was the "green light" for all in the name of development, as the only standard to be respected was market fundamentalism. Huge contingents of hand work of different qualifications were recruited by contractors for construction. According to the Director of the Human Resources Department of Albras,

> During the construction of Albras we have about 5,000 workers at the construction area and 1000 more in the construction of the Village of Cabanos. The main companies that worked in the works were: M. ROSCOE, ESTACON, PARANAPANEMA, MONTREAL, CONFAB, ZANINI, MASCARENHAS, BARBOSA. In the construction of the Village of Cabanos, were the companies below: ESTACON, ENGEPLAN, BETTER, BANDEIRANTES, M. ROCOE, ANDRADE GUTIERREZ (Verbal information).[6]

Companies chose Abaetetuba (a city in the micro-region of Baixo Tocantins) as a place of labor recruitment. It being a polarizer regional center, having reasonable communication service system, trade and transport, as well as the State Secretary of Finance offices, Ministry of Labour, State Department of Transit, Office of the

[5]Interview with Manuel Furtado Silva—councilor in Barcarena, from the period of 1976–1981. In: RIBEIRO, Rosivaldo Furtado. A ideia de progresso em Barcarena. (1980-2002). 2003.65f. Trabalho de Conclusão de Curso de Licenciatura Plena e Bacharelado em História. Universidade Federal do Pará. Belém. 2003.

[6]An interview given in January 2004.

Table 1.4 Constitutional transfers of ICMS, FPM, IPI EXPORT, FUNDEF, IPVA-1995–2003

Years	ICMS transference[1]	FPM transference[1]	IPI[1] EXPORT transference	FUNDEF transference	IPVA transference	Grand Total
1995	5,992,517.54	2,311,972.18	493,983.30	–	–	8,789,473.02
1996	6,953,397.42	2,615,624.24	622,077.44	–	79,176.08	10,270,275.18
1997	6,703,302.59	2,609,913.25	763,638.48	1,138,565.91	113,631.39	11,329,051.62
1998	6,851,736.80	3,180,279.67	705,028.97	2,986,992.58	141,018.09	13,865,056.11
1999	8,476,925.36	4,046,705.63	725,865.99	3,125,263.61	144,433.50	16,519,194.09
2000	10,853,176.00	3,881,009.00	830,777.00	3,656,730.00	153,945.00	19,375,637.00
2001	14,523,007.34	4,605,421.74	973,133.33	4,234,577.73	207,884.30	24,550,024.44
2002	23,276,630.10	5,885,487.71	1,220,103.25	4,565,196.67	264,159.86	35,211,577.59
2003	28,962,949.39	6,134,987.42	1,018,404.98	5,447,462.29	332,864.27	35,195,970.72

Source SEFA/TCU/SEDUC/STN. Prepared by: SEPOF/DIEPI/GEDE. *Note* Nominal values, unless 15% of FUNDEF. Available at: http://www.sepof.pa.gov.br/estatistica/ESTATISTICAS_MUNICIPAIS/Mesorr_Metrop_Belem/Belem/Barcarena.pdf. Access: 08 June 2005

National Department of Roads and Highways. Abaetetuba offered real estate infrastructure able to absorb part of the demand for homes, which increased with the arrival of entrepreneurs.

In the late 1970s and early 1980s, the time of construction of systems of objects needed for the production and distribution of primary aluminum, the dawns of Abaetetuba were marked by intense movement of buses and trucks carrying workers, pedestrians, to the workplaces of Albras/Alunorte. They were the migrant workforce unskilled absorbed in the early stages of implementation of the project's physical structure (deforestation, construction of buildings of the production complex, port, roads and the urban core). According to Fontes (1989) there were workers of Minas Gerais, Bahia, Maranhão, Piauí, Pernambuco, Ceará, and other parts of Brazil; many had participated in the construction of hydroelectric projects such as Tucuruí, Paulo Afonso, Sobradinho, Balbina. Such experiences were added to that of the workers from the region, most of them having had no paid experience before, often attached to their peasant trajectory, with expectations of employers, conducted by moral values, such as honor and loyalty.

To Fontes (1989), the works of the Albras/Alunorte demanded great concentration of workers in Barcarena. These were housed at construction sites, in shacks built by companies, allowing greater control by companies and a disciplinary stance quite rigid, which often allowed abuses, including beatings workers who refused to obey the orders of the heads of sections.

At the end of each phase of the work, a small number of workers (with higher qualification) remained to manage the project phases and inhabited the core of the village of Cabanos. The others had an uncertain fate, preferring to settle in the cities near Barcarena, such as Belém, Ananindeua, Marituba, Abaetetuba, Igarapé-Miri, Mojú, Acará, among others, exacerbating the infrastructure deficit and urban services in these places.

1.2 Locations of Spontaneous Occupation

The implementation of the Albras/Alunorte in the city amounted to 33.46% urbanization rate in 1980 to 47.7% in 1991,[7] but in Barcarena predominantly rural population (Table 1.5). In 2004, more than half of the population lived in rural areas, however, the government prioritized the urban area of the municipality in their management plans. Deficiencies in infrastructure and urban services, which were already huge, consequently, increased due to migration induced by the expropriation of the lands of rural workers and hand contingent of migrant labor attracted to the Albras project.

The fabulous speech that the development of Albras/Alunorte would generate jobs attracted hopeful workers. "The displacement of rural populations

[7]Source: IBGE. Available at: http://www.sepof.pa.gov.br/estatistica/ESTATISTICAS_MUNICIPAIS/Mesorr_Metrop_Belem/Belem/Barcarena.pdf. Access: 08 June 2005.

Table 1.5 Evolution of Barcarena's population IN 1980–2004 period

Years	Total population	Urban population	Rural population
1980	20,021	6,700	13,321
1991	45,946	21,629	24,317
1996	54,259	25,693	28,561
1997	56,084	26,562	29,522
1998	57,622	27,291	30,331
1999	59,162	28,020	31,142
2000	63,268	27,767	35,501
2001	65,385	28,696	36,689
2002	66,913	29,367	37,546
2003	68,604	30,109	38,495
2004	72,441	31,193	40,648

Source IBGE. Prepared by: SEPOF/DIEPI/GEDE. Available at: http://www.sepof.pa.gov.br/estatistica/ESTATISTICAS_MUNICIPAIS/Mesorr_Metrop_Belem/Belem/Barcarena.pdf. Access: 08 June 2005

unemployed, underemployed or surplus to work in state-owned or private enterprises in the Amazon is the basis of the modernization model of the military government to the Amazon" (Ianni 1981, p. 141). So the urbanization rate was high, without improvements in public services, particularly in coastal places, in the countryside and spontaneous occupation of places.

In urban areas there is a deep deficit in the provision of public services: 20.6% of urban households are served with piped water; the sewage network covers only 46.6% of households and garbage collection reaches 87.3%[8] of households. In the health service, the situation is serious, because in 1999, Barcarena had 21 hospitals, nine health centers, two health centers, an outpatient general hospital three mixed units, two offices, two units of family health, a health surveillance unit and a unit unspecialized. In July 2003, this number decreased and the municipality had only 18 of these establishments. In 1999 there were 79 beds, 26 in federal hospitals, 20 in state hospitals, and 33 in the private network. By December 2004 the number of beds remained unchanged[9] and the population increased from 59,162 in 1999 to 72,441 in 2004, which made health care even more precarious.

Population growth demanded expansion of urban public goods and services. In the Urban Plan of Barcarena[10] (except the Rural Colony of Bacuri) are not included areas intended for settlements of migrants or even to the peasants removed to

[8]Available at: http://federativo.bndes.gov.br/destaques/bdg/bdg_mun.asp?idgeo=150130. Access: 10 May 2004.

[9]Source: DATASUS/MS. Available at: www.sepof.pa.gov.br/estatistica/PIB/Analise_dos143_MunicipiosParaense/PIB_Municipa99-02.pdf. Access: 05 May 2005.

[10]In fact, this plan whose preparation was done by a company in São Paulo, "Arquiteto Joaquim & Guedes Associados", should be called Urban Plan of the village of Cabanos, since it covers ONLY

implement the Albras/Alunorte project. The aggravating factor is that neither entrepreneurs nor the municipal government assumes the task of providing infrastructure the spontaneous occupation of places that are built into the design environment.

> The rationality of the planning contrasts with the situations that are emerging in the process of expropriation and reoccupation of space, resulting in the hasty drafting of a settlement program. The situations identified were as follows: a) the social problems caused by urban swelling in Vila do Conde with the arrival of workers from contractors; b) the threat of indemnified families come to allocate in areas that in the future would be expropriated, causing a duplication of benefits; c) the requirement that happens to be made by some of the expropriated families, though diffuse and disorganized fashion, a lot to build a house d) the need to maintain, in areas near the Aluminum Complex, a reserve force work for the "disqualified" services (guard, household, gardening, etc.) in support of the Aluminum Complex and was not expressly contained in the Urban Plan (Maia and Moura 1995, pp. 249–250).

The result of the combination of these factors takes spatial form in the spontaneous occupation of places that do not obey the spatial reordering, architected by the logic of the companies, but witness the increase in regional disparities. In the early stages of formation of these places there is uncertainty in the administrative jurisdiction, because neither the companies nor the local ruling elite are responsible for urbanizing them. The population there lies precariously without any infrastructure, at the mercy of traditional political actions that reproduce contradictorily with the modernization project of public management.

The neighborhood of Laranjal, for example, is characterized as a place of spontaneous occupation. Located between the urban center and living quarters of workers in the Albras/Alunorte, the district set up from the urban settlement policy of the municipality, in the 1980s. Planned by CODEBAR that traced the urban development plan, for the occupation of the expropriated families of the areas defined for the implementation of the Albras/Alunorte project, standardized and built on joint effort buildings were raffled among the participants in 1982.

This construction "in joint effort" system was not based in a solidarity relationship. The settlers did not find reasons to perform an activity that is a responsibility of those who expropriated, placing them in a place devoid of simpler infrastructure. A testimony collected by Maia and Moura (1995) illustrates the situation of forced collective effort:

> At one point, I consider that they have donated, but in another we also sweated to be able to achieve it. I have to me that we shared, we went with the personal help, we worked for nine months, not nine days, it was a struggle, we worked hungry and after what we had, we was eating, after three months things got difficult, because those who had flour could not use it, who had something else could not take to sell, could not because we worked so devotedly, here we could not miss a day of work because they cut us off, they put a rule, as I do not

(Footnote 10 continued)

this part of the territory of Barcarena. Check: Joaquim Guedes & Associados—Plano Urbanístico de Barcarena. São Paulo: Arquiteto Joaquim Guedes & Associados. 1980.

Table 1.6 Sanitation and infrastructure situation of real state in the Laranjal neighborhood

Health infrastructure elements	Regular		Irregular		Absent		Total	
	Absolute	%	Absolute	%	Absolute	%	Absolute	%
Piped water	4	4	56	56	40	40	100	100
Artesian well	59	59	–	–	41	41	100	100
Garbage collection	88	88	12	12	–	–	100	100
Electricity	73	73	27	27	–	–	100	100
Sanitary sewage	–	–	–	–	100	100	100	100

Source Amaral et al. (2002 p. 263)

know how many others, that I do not remember, had some people that lacked twelve days to make flour, sometimes to bring us to eat, because it was difficult, then they already wrote down the fault. (Maia and Moura 1995, p. 251, author's translation).

Forced joint effort to sort the neighborhood did not work because the streets are dotted with wooden houses, masonry or both, built precariously and with lack of infrastructure in much of the neighborhood, as demonstrated by Amaral et al. (2002),[11] these services are either regular or irregular and are affordable for most residents of the neighborhood, with the exception of sewage that is absolutely absent (Table 1.6).

O Novo Bairro is another place that has its origin in the spontaneous occupation. It was formed in 1980, due to both the flows of migrants attracted by the possibility of employment in Albras/Alunorte, as the former besiegers expropriated. The pace of occupations gave up the completion of construction in the company's construction sites, what happened between 1985 and 1990, period in which the extinction of jobs and the reduction in the supply of jobs, being restricted to the guideposts technicians to control tasks. This district covers an area of municipal headquarters, near the PA-151 highway, and is separated from the core planned to the village of Cabanos by the river Mucuruçá. Until 1988, the area served "as garbage disposal and there the pauperized population tried to get utensils and food necessary for their reproduction" (Teixeira and Oliveira 1995, p. 264).

In Bairro Novo, although it is a spontaneous occupation area, the local ruling elite have spread their territoriality, understood here as the "set of strategies, actions used to establish this power, maintain it and reinforce it" (Gomes 2002, p. 12). The interest of the elite to urbanize this part of the territory in Barcarena is explained because a part of the district belonged to the Dias Melo family that developed the production of flour and charcoal for trade. The municipal government paid compensation to the family for part of the invaded land then opened streets and donated lots.

The Bairro Novo has become an area of expansion of the municipal headquarters, where are the administrative and state social policy buildings, such as the Welfare Foundation of Para and the Cultural Center of the City where fairs,

[11]The survey encompassed 100 households and was conducted in the years 2000–2001.

festivals, and other activities are organized, in addition to having numerous pharmacies and sales, through which the people find a means of survival.

The Bairro Pioneiro is a place with spontaneous occupation traits. Designed by CODEBAR to be a temporary camp where would inhabit the workers of the contractors after the works would entail a commercial center of the urban core of the village of Cabanos. According to an interview given by a member of the association of traders at Baitto Pioneiro:

> There was no Pioneiro neighborhood, so there were some points like bus station is, cinema, a few little things, what happened. In the commercial project for the Vila Cabanos, in the beginning of it, the commercial center would be there where the bus station is, but suddenly it was directed to Vila dos Cabanos, and here was abandoned. CODEBAR canceled the sale of lots for low-income people. What happened was that people began to cluster together: in a house lived 3, 4, 5 or 6, the day came the need arose the invasion, which was quick and took care of all the Pioneer neighborhood. (Verbal information).[12]

The Bairro Pioneiro that was built distanced itself from what was planned. At the end of the works, the barracks were turned off and the area practically abandoned. It came back to life in 1996 with the occupations.

> When you asked about the invasion, what was this invasion? People came from Abaetetuba, Moju, Iguarape-miri and came to work here and later returned to their homes, but the car turned on the road, died, broke its leg, all this occurred at the beginning of the project. In the implementation of Albras/Alunorte Project there was not a concern to charge by the government. Let's say, if the Albras will make the building core, it is necessary to build a popular area in case of the Bairro Pioneiro, then the invasion came (Verbal information).[13]

Classified as an illegal and irregular area of occupancy, neither entrepreneurs of Albras/Alunorte project nor the rulers were responsible for the urbanization of the neighborhood. According to the Director of the Community Center neighborhood we interviewed:

> We have no business to participate, they do not help, they never helped, said they had a budgeted allowance for basic sanitation in Bairro Pioneiro, so far it hasn't arrived, there never was a shareholding of Albras/Alunorte. We know that there is a company here, because the Bairro Pioneiro was made from a raid and who was in charge of this area was CODEBAR because this is the land of the union, then who was in charge here was CODEBAR, but really, this company which is CODEBAR, which means Barcarena development company is not developing anything, it just came to disrupt the situation here in Barcarena (Verbal information).[14]

The action of the Community Center of Bairro Pioneiro has been intense. Residents are organized to have access to infrastructure services, especially on the spontaneous occupation of the neighborhood. According to our interviewee, a manager of this community center:

---

[12]An interview given in March 2005.

[13]An interview given in March 2005.

[14]An interview given in March 2005.

> We here at Bairro Pioneiro, we were never sought by the city to ask or then say where is that income should be applied. Let's say, basic sanitation here in Bairro Pioneiro is zero sanitation and street lighting. We even invite councilors to attend a meeting to collect on this, they came here and promised they would do (something)... so far nothing (...) We even have a recorded tape, an article showing the situation we're in, and we have already been charged in the sub-prefecture with the deputy mayor, because it is on that side. Because it was divided, it was always Murucupi district, now the responsibility was divided, here, who takes care is Abaeté, who is the deputy mayor, but he can only do something if the mayor authorizes. (Verbal information).[15]

The discontent in the narrative indicates that our interviewees recognize the unequal production of space, witnessing the unfair distribution of social wealth, as reported in a leader of the community center in São Francisco that:

> The wealth of Barcarena are here on the side, São Francisco, Laranjal, Pioneiro, Vila de Cabanos, Vila do Conde and Itupanema, because of the industrial pole. And the money is invested in the headquarters of the city, we think they care more about the headquarters of the city. And here, where wealth is generated, it is that way (Verbal information).[16]

The residents perceive the unequal territorial distribution of public investments and recognize fragmentation and hierarchy between different places within the municipality. They are angry for not taking advantage of the wealth produced in the territory of Barcarena, leaving them to use the territory as a shelter, where they live and spatial basis for proliferation of welfare practices and political subordination to municipal managers.

1.3 A Fabulous Place Planned: The Village of Cabanos

The conservative actions system included the planned village of Cabanos. In Para, beyond this village, we have the residential villages in Tucurui, linked to Eletronorte and the Carajas' urban center, connected to the company Vale do Rio Doce. These places, according to Trindade Jr. and Rocha (2002): (a) act as a sort of extension of the company's production line to which they are attached; (b) by their nature, design and density techniques, they tend to deny the regional urbanization patterns, defined throughout the production of Amazonian space process; (c) characterize a new form of management of local and regional space, given the relative economic and political autonomy that are invested; given that, as corporate headquarters, centralizing decisions have financial resources and concentrate most of the qualified personnel; (d) although in some cases present in open cities, the urban design that characterizes closed nature, end up defining them as true urban enclaves, able to ensure the activities of the company and control of the labor force.

---

[15]An interview given in March 2005.

[16]An interview given in March 2005.

In the planning period of the villas, the political action system and the normative densities mobilized by the Brazilian government direct the public fund for reproduction of capital production conditions. We see this when we consider the amounts invested and funding sources for construction of these cores. For the urban center of the village of Cabanos:

> According to estimates of the Interministerial Commission (Brazil 1977), the total cost of the urban core, in dollars as of December 1977 would be U$ 137,243,000, distributed between: works of physical and social infrastructure (U$ 46,732,000) to be defrayed by specific funds, such as the Program of National Integration (PIN) and the Strategic Areas Development Fund (FDAE), these resources are repayable; construction of camps intended to serve as housing for about 10 thousand workers provided for all the works (U$ 25.29 million), to be funded with resources from ALBRAS and Alunorte, in charge of CONSOAL task; construction of 1922 apartments for singles and 2885 houses for the population directly linked to the project (U$ 65,221,000), to be financed borrowers (residents) with funds from ALBRAS and ALUNORTE, construction also in charge of CONSOAL.
>
> Not included in these costs is the construction of 4329 homes and 2887 apartments for those single, for the other inhabitants of the core, to be financed and built by private enterprises. They were also provided U$ 11,683,000 to be applied in the urban areas adjacent to the project (Lobo 1996, p. 117, author's translation).

The village of Cabanos was conceived as a modern place. The Brazilian state was responsible for financing and building this logistical support operation of the Albras/Alunorte complex. Also according to Lobo (1996), to manage and perform the works of the new urban core and the area adjacent to the project, in addition to works in neighboring municipalities,

> It was created by the aforesaid Administrative Rule [Ordinance No. 061 SEPLAN-PR of 04/12/1982], at the suggestion of the Special Group, the Development Company of Barcarena (CODEBAR), pursuant to Law No. 6,665, of July 3, 1979 with a capital of US $ 26,051,400, divided into 70,000 common shares. The company also received an additional appropriation of $ 948,700 [...] the resources of additional title for your installation. The borrowing operations carried out by the Company have guarantees from the Federal Government and are free of Union competence of taxes. The Codebar replaced SUDAM in local coordination task of the Support Program for Barcarena Industrial Complex.
>
> CODEBAR took the form of a public company with majority interest, however transitory, the Union and the state and municipal governments, providing its creation law that the shares owned by Union should in future be transferred to the City of Barcarena (Idem, p. 116–117, author's translation).

In the Urban Plan of Barcarena, the village of Cabanos is presented as an "open city" accessible to "all who seek":

> The future microregional pole condition it is supposed to the new core, relies on its undeniable character of the city open to all who seek, as a condition, contrary to popular belief, of quality. The project of social areas has been drawn up on a program of open activities, alternatives to the entire population, with a view to mitigating the effects of isolation and identity references, attacking the typical pathological relationship of this situation (Sudam 1980, p. 62, author's translation).

Following the development pole idea of III Amazon Development Plan, this "open city" would assume the function of integrating the micro-region where it operates. It would be able to have a population between 40,000 (minimum hypothesis) and 70,000 (maximum hypothesis) inhabitants by the year 1988, but did not summarize the logistical support condition of Albras/Alunorte and the industrial district. According to Trindade Jr. and Chagas (2002), in infrastructure would be a city with many possibilities of inclusion and integration, it would occupy a prominent position in relation to nearby cities, a kind of central location.

But this fabulous plan was off. Changes in regional development policies were undertaken to consolidate the industrial district of Barcarena, as provided for in the original design of the Albras/Alunorte. The implementation of the development

Table 1.7 Comparison between the urban plan of Barcarena and produced space—public transport

Urban plan of Barcarena	Produced space
Ferry boat: waterway connection between the project area and Belém, with exit of Barcarena terminal	Ferry boat: there are only two lines of ferries controlled by five companies. One is the city's connection to Belém; the other connects the core to the headquarters of the city
Bus: intra urban transport system; highlighting the advantages it offers: small size, flexibility and acceptance	Bus: meets a small portion of the urban core. There is a kind of public transportation, which caters exclusively to the Albras/Alunorte and connecting the core to the factories
Trolley-bus: transport to be deployed when the nucleus count on a population of 50,000	Trolleybus: nonexistent

Source Trindade Jr. and Chagas (2002 p. 222)

Table 1.8 Comparison between the urban plan of Barcarena and produced space—education equipment

Urban plan of Barcarena		Produced space	
Demand to be met	Estimate (number of students)	Attended demand	Numbers
Infant/day care	8310	Child education	800
Preschool	5620	Elementary School	2451
Elementary school (1st cycle)	6895	High school	914
Elementary school (2nd cycle)	6326	Supplementary education	494
High school, professional and supplementary	32,839	Higher education	Nonexistent
Higher education	702		

Source Trindade Jr. and Chagas (2002, p. 217)

Table 1.9 Comparison between the urban plan of Barcarena and produced space—social promotion, sport and leisure

Urban plan of Barcarena	Produced space
Reception and referral center	Cabana club
Cultural spaces (libraries, auditoriums, theaters, cinemas and places for meetings)	Club 1—bars and restaurants
Sports centers	Deactivated cinema
Four spaces for activities not designed: play area or recreation	Video rental stores
Four spaces for spontaneous soccer practice	–
Four or more spaces for institutions such as SESI, LBA, etc.	–
Four spaces for other forms of recreation (near the beaches)	–
Spaces for theaters, bars, cafes, discos, amusement park (commercial and central centers)	–

Source Trindade Jr. and Chagas (2002 p. 220)

Table 1.10 Comparison between the urban plan of Barcarena and produced space—health equipment

Urban plan of Barcarena	Produced space
General hospital: 195 beds	Hospital: 50 beds
Pediatrics: 67 beds	Medical observation: 12 beds
Obstetrics: 28 beds	Emergency: 1 bed
Medical clinic: 60 beds	Hospitalization: 37 beds
Surgical and obstetrical center: 7 rooms	–
Nursery: 30 berths	–
Auxiliary services: 3 rooms	–
Complementary services: laboratory, radiology, dietary kitchen, laundry, warehouse and pharmacy	Services: emergency room and outpatient clinic, operating room, nursery, hospitalization sector, hospital pharmacy, nutrition and dietetics service, laundry, medical and dental offices, physical therapy, general examinations, medicine and occupational health, preventive medicine, morgue
Sanitation (water control, sewer, garbage), public nutrition, control of infectious and parasitic diseases, morgue	

Source Trindade Jr. and Chagas (2002, p. 225)

plan of the village of Cabanos fell to places to meet primarily to demand and command of ventures.

In 2002, a population of about 8000 inhabitants lived in the village of Cabanos, fewer than shown in the Urban Plan. As shown in Tables 1.7, 1.8, 1.9 and 1.10, the territorial configuration proposal is quite different from what we find in the Plan.

The project of a city open and integrated with other municipalities in the region did not take place. In fact, the core of the village of Cabanos summed up the residential locus function, where the architecture and quality of hierarchical housing (according to the salary range) exemplify the uneven social and territorial division of labor, very different from that proposed the project: to mix "carefully all income levels, avoiding the segregation of lower income populations in peripheral or external slums, cast away from urban area and its basic infrastructure" (Sudam 1980, p. 70, author's translation).

Due to changes in the orientation of regional development policies and plans of privatization of federal and state governments, the municipality of Barcarena and therefore the village of Cabanos are no longer concern and responsibility of companies. The privatization of Vale do Rio Doce Company (CVRD) and reforms in the allocation of public funds influenced the policies pursued by Albras/Alunorte. Companies gradually transferred responsibility for the maintenance and provision of services and infrastructure that remained to the city. In the legal framework, when the Municipal Organic Law transferred virtually all the functions of control, supervision and implementation of the village to the city of Barcarena, redefined the management responsibility of competence of this space. The city was gradually taking over the maintenance and expansion of urban services and infrastructure in place allocated.

The residents of the village of Cabanos were not indifferent to the ongoing land management transfer. Among the village residents who once despised the idea of being of Barcarena and not resorted to the mayor and council members to demand improvement and expansion in urban goods and services, it strengthened the practice of resorting to local ruling elite to resolve place issues. Including community leaders that came to enjoy political power, often fielding candidates for councilor.

There was, therefore, a transition in the management of the village. The privatization of Vale do Rio Doce Company (CVRD) emptied the political power of the Development Company of Barcarena (CODEBAR) on the village of Cabanos. At the time of its conception, the one who should fund and build the place was the city. But CODEBAR took over the management of the project and has since limited the participation of state and local governments in coordinating the works, on the grounds of administrative capacity and debt, both would have to drive forward the process to the amount of investment required by the project.

Today we have a new process of transfer of responsibility, now from CODEBAR to the city of Barcarena.

> What codebar has effectively done is to provide technical support for actions in conjunction with the city, as well as the marketing of lots available in the village, analysis of architectural projects since it is without resources for investments in the purchase of materials and equipment (Rodrigues 1998, p. 26, author's translation).

The tax collection and increased municipal revenues provided by Albras/Alunorte created budgetary conditions for the effective transfer of responsibilities. The fiscal and budgetary health of the city should translate into quality of life for the population, but high numbers allocated to the multi-annual plans contrast with the poor infrastructure and public services. This has motivated the political elite centered on the village of Cabanos proposing emancipation of political and administrative seat of the municipality of Barcarena.

Thus the setback of privatization policies Albras/Alunorte and CODEBAR left for public bodies the burden of managing urban problems. On the other hand, the conservative political elite took advantage of this moment of transformation to garner political power. The city is reduced, then the instrument now an economic organization, according to which urban needs are inventoried and guided; now with the organizer pretext of public space, political power dissolves the social organism, the subordinate class organizations and urban services to its decisions and needs, more often electioneering and conservative.

References

Amaral MDB et al. (2002) A Outra Face da Cidade-empresa: os assentamentos urbanos na área de influência do projeto Albrás-Alunorte. In: Trindade SC Jr, de Rocha GM (Orgs). Cidade e Empresa na Amazônia. Gestão do Território e Desenvolvimento Local. Belém. Paka-Tatu, pp 237–271

Bobbio N et al. (1991) Dicionário de Política. 3ª Edição. Editora da UNB, Brasília

Becker B, Egler CAG (1994) Brasil. Uma Nova Potência Regional na Economia-Mundo. 2ª Edição. Bertrand Brasil, Rio de Janeiro

da Costa Gomes PC (2002) A Condição Urbana. Ensaios de Geopolítica da Cidade. Rio de Janeiro. Bertrand Brasil

de Oliveira AU (1993) Amazônia. Monopólio, Expropriação e Conflitos. 4ª Edição. Papirus, Campinas (SP)

de Oliveira F (1998) Os Direitos do Antivalor. A economia política da hegemonia imperfeita. Editora Vozes, Petrópolis (RJ)

de Vasconcelos AM (1996) Transformações Sociais e Lutas Cotidianas no Laranjal em Barcarena. Relatório Final de Pesquisa. Belém. UNAMA

Dimenstein G (1995) Meninas da Noite. A Prostituição de Meninas-Escravas no Brasil. São Paulo. Editora Ática

Ianni O (1981) A Ditadura do Grande Capital. Civilização Brasileira, Rio de Janeiro

Ianni O (1986) Ditadura e Agricultura. 2ª Edição. Civilização Brasileira, Rio de Janeiro

Lobo MAA (1996) Estado e Capital Transnacional na Amazônia: o caso da Albrás-Alunorte. UFPA/NAEA, Belém

Maia MLS, Moura EAF (1995) Da Farinha ao Alumínio: os caminhos da modernização na Amazônia. In: CASTRO, Edna; MOURA, Edila A.Ferreira; MAIA Maria Lúcia Sá. (Orgs). Industrialização e Grandes Projetos. Desorganização e Reorganização do Espaço. Belém. Editora Universitária.UFPª, pp 223–260

Nahum JS (1999) A Amazônia dos PDAs: uma palavra mágica? 1999. [s/r]. Dissertação de Mestrado em Planejamento do Desenvolvimento do Núcleo de Altos Estudos Amazônicos da Universidade Federal do Pará. Belém

Pinto LF (1982) Carajás, o ataque ao coração da Amazônia. 2ª Edição Ampliada. Rio de Janeiro. Editora Marco Zero

PONTE (1985) Tereza Ximenses. Mutirão, campo e cidade. São Paulo

Rodrigues RM (1998) Vila dos Cabanos: o período de transição. Relações Urbanas e Perspectivas de Gestão. 1998.60f. Monografia de Especialização do XIII Programa Internacional de Treinamento em Projetos de Desenvolvimento de Áreas Amazônicas. NAEA/UFPª

Santos M (1991) Metamorfoses do Espaço Habitado. 2ª Edição. Editora HUCITEC, São Paulo

Santos M (1999) A Natureza do Espaço. Técnica e Tempo. Razão e Emoção. 3ª Edição. Hucitec, São Paulo

Santos M (2000) Por uma Outra Globalização. Do pensamento único à consciência universal. 2ª Edição. Editora Record, Rio de Janeiro

Santos M (2005) Da Totalidade ao Lugar. Edusp, São Paulo

Sudam (1980) Plano Urbanístico de Barcarena, vol 2. Guedes e Associados, São Paulo

Teixeira JB, Oliveira E (1995) A Política Social da Albrás Destinada às Populações Ribeirinhas. In: Castro E, Moura EAF, Maia MLS (Orgs) Industrialização e Grandes Projetos. Desorganização e Reorganização do Espaço. Belém. Editora Universitária. UFPª, pp 261–284

Trindade SC Jr, Chagas CAN (2002) A *Company town* do alumínio: concepção e práticas espaciais. In: Trindade SC Jr, de Rocha GM (Orgs) Cidade e Empresa na Amazônia. Gestão do Território e Desenvolvimento Local. Belém. Paka-Tatu, pp 195–235

Chapter 2
The Use of the Territory as a Resource by the Ruling Elite of Barcarena

Abstract In this chapter I reveal that, with the exception of the industrial complex area, the rest of the territory of Barcarena, including the Village of Cabanos, is controlled by a system of conservative actions, administrative centralization promoter, lack of transparency, and absence of participation of civil society in the preparation of administrative planning. In sum, old production practices of territorial inequalities. Also, before upgrading vectors, especially those who come to regulate the processes of decentralization and administrative democratization, the mayor, councilors, and secretaries decide how to implement them in a way that suits their particular interests. Ensuere in Barcarena that which happens in all of Brazil.

Keywords System of conservative actions · Administrative centralization · Decentralization

The first chapter held that a stock system reproduces conservatively every modernization process, showing that the territorial reordering of Barcarena is closely associated with this system's actions. In this reorganization, the Albras/Alunorte is an agent that uses the territory as a resource to ensure the effectiveness of capital reproduction process. The modernizations introduced due to the installation of complex Albras/Alunorte does not set to deny the power of the local ruler elite, nor minimize the centralizing power of this elite. On the contrary, they are allied solidarity and organizationally to use more effectively the resources of the place. Faced with new vectors of administrative modernization, for example, standards of public administration, the whole territory is controlled by this system of actions that conserve traditionally actions centralizing policies, intransparent and suppressive of initiatives and standards that want to ensure popular participation in planning and administration of public affairs.

In addition, on vectors of administrative modernization, the mayor, councilors, and secretaries decide to implement them in order to meet their particular interests, restoring at Barcarena something recurrent in Brazil, especially with regard to the new standards of public administration, which seek to regulate public spending and traditional political action, that is, the old despotism practices, welfarism.

J.S. Nahum, *Modernization and Political Actions in the Brazilian Amazon*, SpringerBriefs in Latin American Studies, DOI 10.1007/978-3-319-58030-2_2

Such relations between the modern and the traditional have in Barcarena played what J.S. Martins called *the power of the delay*, stating that "the social and political transformations are slow, and are not based on sharp and sudden social, cultural, economic and institutional ruptures. *The new comes as an offshoot of the old*" (Martins 1999, p. 30, author's translation), which allows us to adjectivize as fabulous this project of modernization, because it does not break with the old and traditional conservative political practices.

2.1 The Conservative Actions System

The Barcarena city is a place that has been planned and managed conservatively by the local elite in the interests of economy class and politically dominant. We understand as Miliband (1999) that

> a ruling class in any class society is recognized due to its effective control over three main sources of domination: the means of production, where control may involve (and usually involved) the ownership of companies, but does not necessarily need to do it; the means of administration and enforcement of the State; and the main means to establish communication and consensus (Miliband 1999, p. 476, author's translation).

The elite in the ruling class is recruited accordingly, all elite is ruling class, even though the reverse is not true. Although the ruling elite defend particular interests of the class, this does not eliminate conflicts of interest between them. Interests that are manifested mainly in electoral periods.

There is a conflict, a dispute, a dispute between the representatives of the political elite, it aimed to expand influence and prestige therefore commanding the municipal administrative structure. Individual motivations, even causing conflicts do not threaten the defense of common interests to the ruling class. Therefore, the third term of Laurival Cunha Campos ended with the appointment of Wandick Gutierrez as mayoral candidate, with the condition that he keeps the administrative organization to inherit.

A traditional political style, represented by two family names, which act in a hegemonic way in Barcarena territorial management, whether inside or outside the administrative structure. In this style governed: Laurival Campos Cunha and Laurival Magno Cunha (father and son) in the periods of 1963–1967; 1983–1989; 1993–1996; 2001–2004 and re-elected in 2004; and Wandick Gutierrez during the 1989–1993 administrations, from 1996 to 2000, being defeated in the election of 2004. Despite alternating names, but the conservative political actions of class power that are structured to withstand threatening modernizations to conservative way to manage public affairs remain.

These same conflicts within the class and the same associations that guarantee the conservation of political power class can be exemplified by the analysis of the electoral process in 2000. At that time, due to the death of former mayor Laurival Cunha Campos, the elite sought a replacement candidate who could no longer be

Wandick Gutierrez, given his unpopularity. Among the candidates: Father Adamor Lima (PT), João Evangelista Vaz (PSL), Adinaldo Souza de Oliveira (PV), Maurilo da Costa Gomes (PFL), João Carlos dos Santos Dias (PMN), and Laurival Magno Cunha (PMDB). The political elite supported the latter and their application by the Party of the Brazilian Democratic Movement (PMDB) with the support of state party leaders, as Jader Barbalho, Priante, Elcione Barbalho and local traditional families, such as Rodrigues, from where came the former mayor José Pinheiro Rodrigues. Laurival Magno Cunha, son of Laurival Campos Cunha, was little known, had no traditional political image, he was not a candidate, did not attend local cultural fairs, political rallies, and even corporate openings in Barcarena.

Mr. João Carlos dos Santos Dias, candidate for the Party of National Mobilization (PMN) is a local businessman in the grocery business and had until then a very expressive political history, he was deputy mayor of Wandick Gutierrez management from 1989 to 1993. The Santos family has supported for a long time the political group of Wandick Gutierrez, but withdrew this support when the mayor lost its popularity due to an expressionless administration.

The priest Adamor Lima, candidate of the Workers Party (PT), does not belong nor represents the traditional local families. Born in Abaetetuba, Adamor Lima works with social movements, especially of rural workers. Invited by PT in April 2000 to be candidate, he accepted the invitation even without the support of the Catholic Church, contrary to electoral involvement of its leaders. Without the support of the Church, without coalition with other parties and unassociated with traditional families, his political campaign was not significant. But his poll showed that a significant part of the population did not agree with the maintenance of the power of traditional families via relay positions in municipal governments.

The campaign polarized between João Carlos dos Santos Dias and Laurival Magno Cunha representatives of the two economically strong local groups, traditional family business in the city. This time of the campaign, there was a clash between the two groups, and this is intensified with the result of the calculation, assuming violently and real conflict between the supporters of Laurival Magno Cunha (which won the 2000 election with 38.84% of votes valid) and João Carlos dos Santos Dias (who got a strong vote of 35.98% of the votes[1]).

Around the country: VOTERS BURN FORUM

A protest against the outcome of the election for mayor of Barcarena, a city located 25 km north from Belém (PA), resulted in the invasion, depredation and fire in the Forum and the headquarters of the State Prosecutor in the early hours of yesterday. Computers and diskettes were totally destroyed and thrown to the ground. More than 200 men from the military police were sent from the capital to ensure order. Ten people were arrested and two shot during the riots. One of the victims, Armando Dias Pimentel, was taken in serious condition to the Municipal Emergency in Belém. The district judge, Carlos Flexa, asked for

[1]The priest Adamor Lima got 22.27 of the votes; John the Evangelist Vaz got 2.15, and Adinaldo de Souza Oliveira got 0.83. Source: www.tse.gov.br. Accessed December 1, 2005.

police protection. His house is surrounded by military police. Two prosecutors had to leave the city, escorted. The fire in the electoral office of the facility was only put out in the early morning[2].

In the words of the president of a community center in the city of Barcarena,

> Most people said they burned the forum because they believed many people would vote for João Carlos. Wherever you went the result as João Carlos, we did research in several places the result was always João Carlos, and when they arrived at the ballot box they do not know what happened, it was a very strange thing because in the vote count, Laurivalzinho pulled ahead. So in people's minds the election was rigged (Verbal information).[3]

However, such differences in the political elite are ephemeral. The reconciliation took place in the 2004 municipal elections, a time when many believed that Mr. João Carlos would run in opposition to Laurival Magno Cunha. The population showed to be rather disappointed. In an interview with our interlocutor they tell us,

> I even feared that (it would happen) again, and even justice feared so much that they sent another judge, there was much policing because they feared the same thing. Because people were a little angry with João Carlos who was candidate for mayor and at the time was against Laurivalzinho and was now on his side. So people who were waiting for him to come to try again were disappointed because he went to the side of the mayor. So people were very divided. Laurivalzinho won because the other candidates allied to him. I voted in Laurivalzinho because I did not see another candidate who would replace him (Verbal information).[4]

The intransigent defense of private interests is what unifies these groups of traditional families and feed their relay in charge of the government. Despite the modern standards of public management, the growth of municipal tax revenue resulting from the installation of a modern industrial complex in the city and on the most urgent collective goods needs for the local population, *the traditional political elite keep their old political practices.*

2.1.1 The Dispatches Policy

The dispatches policy is the mayor and aldermen strategy to engage with the public. In election years the practice of orders turns the campaigns in a show whose outcome is determined by the power of political marketing to make the applicant an acceptable goods through the numerous visits to more distant locations, distribution of gifts, vows import of barcarenenses voters living in other municipalities, the rallies that turn into true parties with presentation of folk musicians, prize sweepstakes. All supports to win the vote of voters who pay little attention to the speeches of the candidates, who are interested in having their individual demands addressed.

---

[2]Source: http://an.uol.com.br/2000/out/04/0pot.htm. Acessed June 13, 2005.

[3]An interview given in March 2005.

[4]An interview given in March 2005.

In election campaigns of Barcarena—like all of Brazil—the election period, there are candidates with an air of humility, visiting everywhere, sympathizing with the demands of all, celebrating the various agreements. According to our interviewee, a director of the community center of the Laranjal Neighborhood:

> In the campaign season they promised heaven and earth. At least that commitment made in the campaign would have to meet. For example, at the time of the campaign, the mayor came here with us, said that he would pave the streets, create basic sanitation, have water. And none of that is happening. So we see a lack of commitment, lack of interest even by the community. Also because our city today has a very large population, so it was not supposed for us to be living the way we are living today. I do not know if you noticed some streets there can not even move through because the garbage is closing them.
>
> I ask: Why is that?
>
> As I said to you just now, it is the lack of same commitment. The "man" comes here, offers, promises several things that will really benefit the community, then okay. Then everyone stays [...] in fact we were cheated again, because everyone trusted he would do, and it is not what is happening (Verbal information).[5]

Once sworn in, the so-called politicians do not keep their promises. This probably explains the level of mistrust and rejection that people have of politics. This is because in the popular imagination, politicians are the ones who transform public goods into private benefits and that management involves suiting individual requests. Why voters flock to the portal that gives access to the offices of councilors, demanding the promises made during the campaign, but predominantly making requests to solve individual problems. During the year, the political activity of the councilor is to dispatch things. In the words of a former alderman:

> I often say that the alderman's activity begins in January. People hit the councilor's door for? They want the Councilman to intercede with the mayor to get a work contract. Also, around the end of January when "the guy" has no money, he resorts to the councilor to pay for his gas; then starts a pressure in the alderman for these things. People who will ride carnival group in February also have begun to look for you in January. When it comes to February when the school starts, the issue are school supplies. So the neediest parents and people who are most in need and others until they have no need, but has Councilman knowledge, already looking Councilman because they want school supplies, uniforms, shoes.
>
> When comes the month of March is then one last time for the student to enter the school uniform. So when I see that parents run telling me, "Oh councilor, if my child is not with the uniform he could not enter the school because the teacher said he will not allow more because it has to be uniform". Then the councilor has to give. Out the day the natural Alderman day that is one issue of the medicine, the examination, the passage to Belém, that is daily.
>
> When it comes to Easter, all schools make the Easter party, all classes, classrooms promote party for students. And the teacher has to take the cake and it is here in the Alderman door he comes knocking to ask for help for these Easter events are held.

[5]An interview given in March 2005.

> When comes the month of May, the second Sunday of May, Mother's Day party in all communities, all schools have Mother's Day party. Is that pressure on the councilor to cover and support all these parties.
>
> There comes the month of June is the worst month for Alderman in Barcarena because it's folkloric group everywhere.
>
> Then you think that July is recessed the chamber for Alderman will play. It is usually in the month of July that people go on vacation, want to travel, want to visit a relative in another city then starts to pressure because of passage. Requests always personally, nothing collective, nothing communitarian.
>
> There begins the month of August, starts the second Sunday with Father's Day. It is the same greatness of Mother's Day, but still have programming throughout the city for Father's Day.
>
> When comes the month of September begins with the September 7, then the father arrives so, "but councilman if my child does not parade it will not make a point that the teacher said he would give him [...] you have to give clothes to the person parade, help the school to assemble band".
>
> Here comes October, when it arrives in October there soon coming Children's Day, on the 12th is the same problem in relation to schools, all schools send letters, I've got to get more than 100 a month in relation to schools asking requests. Then has the issue of Cereus in Belém, our people are very religious, is very believer, leads to faith, it is very present in the life of our people. So people want to go to Círio's Nazareth and is passing, is the clothing of the child that "man" does not have to buy is the promise he made to you to win the election and he has to pay now there with Santa, he undertook around the candle to go there because he made a promise in your name but you wouldst not win the election, then you have to pay for that expense, no way.
>
> When it comes November, starts November 2, All Souls. It candle to all the people, the family is there wanting candles, wanting to lighten their loved who have died, people who have to leave the city, people who come to the city and not go home again, and you have to give passage back. When comes the second Sunday of November is the candle of Barcarena. There is the councilman being charged to sponsor video of Cereus and programming of Cereus, and also the cost increase in the house of everybody because it is tradition that meeting of Cereus on Sunday. Also in October is the day of teachers. November is the month on the second Sunday cereus (Verbal information).[6]

In the logic of the rulers, legislate and administer are synonymous of dispatching, equating them to the dispatcher figure, as defined by Da Matta (2001),

> The figure of the dispatcher, the specialist that contacts government offices to obtain documents generally involve the confusion I mentioned before lines, describing the 'way' they do it. The dispatcher, as a sociological figure, can only be seen in its huge importance when again we realize this huge brazilian difficulty of joining the law with everyday social reality. Thus, the dispatcher looks more like a godfather. Just as an employer must provide employment and good working conditions for its employees, the dispatcher must guide their clients through the narrow and dangerous intricacies of government offices, causing them to follow the right path (pp. 102–103, author's translation).

It is understood by dispatching the mayor's reception, be it in his home or in his office, of the members of the direct administration, and especially the people. In the

---

[6]An interview given in March 2005.

territory of Barcarena, things are dispatched as if the municipal public institution, its structures and resources were the private domain of the manager. The public management takes personal connotation, as if the manager was managing his home, uniting, as pointed by Da Matta (2001) the law with everyday social reality.

In this political practice as in patrimonialism examined by Faoro (1991), dissolves any boundary between the interests of the public institution and its interests. The corollary of this is that the common good becomes a fallacy, as well as the public space, the public sphere, public management, and citizenship in the territory become chimeras. Instead we are witnessing the decline of citizenship as

> social pact simultaneously established as a relationship belonging to a group and belonging to a territory. This voluntary agreement is formal and aims to ensure the rights and duties of each individual. Cohabitation of these individuals is well over a space that is also the subject of a formal agreement, which establishes limits, indicates uses, sets parameters and signals the interdictions. This type of standardized space is the matrix of public space and the main locus of reproduction of collective life, and all social action seeking to subvert the existence of this space or change the statute is necessarily redefinidora the terms and corresponds to a decrease of the initial contract based citizenship, retreat that is both the institutionalization of social practices that make up a democratic framework of life and citizen as physical, arrangement material limits and eligible actions (Gomes 2001, pp. 173–174, author's translation).

When the municipal government does not recognize those limits, does not have a public space, where according to Arendt (2003), the prevailing political action that takes place in recognizing the other as a condition of existence of the self, and vice versa; place where the prevailing *plurality*, *the common interest*, which is neither individual nor social, is *public*. Thus, we have the action of public policy manager, what you have is the action of the ruling elite leading social assistance policies, using the territory as a resource to preserve their interests.

To Oliveira (2004, p. 41), "the so-called 'social policy' would be 'focused and not necessarily public', consisting of a poverty functionalization instrument". Hegemonic agents of the political use that part of the municipality where lies the poor, as a spatial foundation for their welfare actions. We can even say that the maintenance of poverty in these places has the function to ensure the continuity of the ruling elite, i.e., those territories justify social policies and prevent the implementation of public policies. It's a vicious circle.

Control mechanisms and oversight of public management are safe, work according to the system of conservative actions, therefore managers behave as public property owners, using municipal resources according to their own convictions and interests. Political activity is reduced to the management of individual interests. This allows them, for example, not to prioritize the collection of municipal taxes, especially the Building urban land tax (IPTU) which also constitutes clientelist practice.

In 2000, IPTU revenue was projected at R$ 876,370.00 and a total of R$ 921,583.00 (Table 2.1). In 2001, there was a prediction of raising R$ 1,000,000.00, projecting a default of 50 and 2% of exemptions, was raised R$ 743,895.00. The estimated revenue for 2002 was 40% of what was predicted in the previous year.

Table 2.1 Statement of evolution in IPTU Levy in Barcarena, 2000–2003

Exercise	Forecast	Levy	Exemption	Bad debt
2000	876,370.00	921,583.00	–[a]	–[a]
2001	1,000,000.00	743,895.00	2%	50%
2002	400,000.00	845,906.00	2%	30%

Source Budget Guidelines Law for 2001, 2002, and 2003. Organized by the author. No data available in the documents

Although it is known the preparation of the Budget Guidelines Law for 2002, the IPTU revenue had reached almost designed for the 2000 financial year.

In Barcarena territory, a period was lived in which the arrival of technical engineering systems and modern management administration has been accompanied by the strengthening of traditional policies, particularly patrimonialist which uses the public power for the benefit of private interests and to the detriment of the interest and population's needs. The strategy of orders takes an institutionalized function of a mode of action and the population is familiar with such an institution, for the welfare policy of the orders is more widespread, used and known that their own citizens' rights set out in the Letters Federal, State and Municipal Constitutions.

The welfare actions enlivened by politics thrive in Barcarena, in the territories of spontaneous occupation of the poor, which we have already referred to. Are

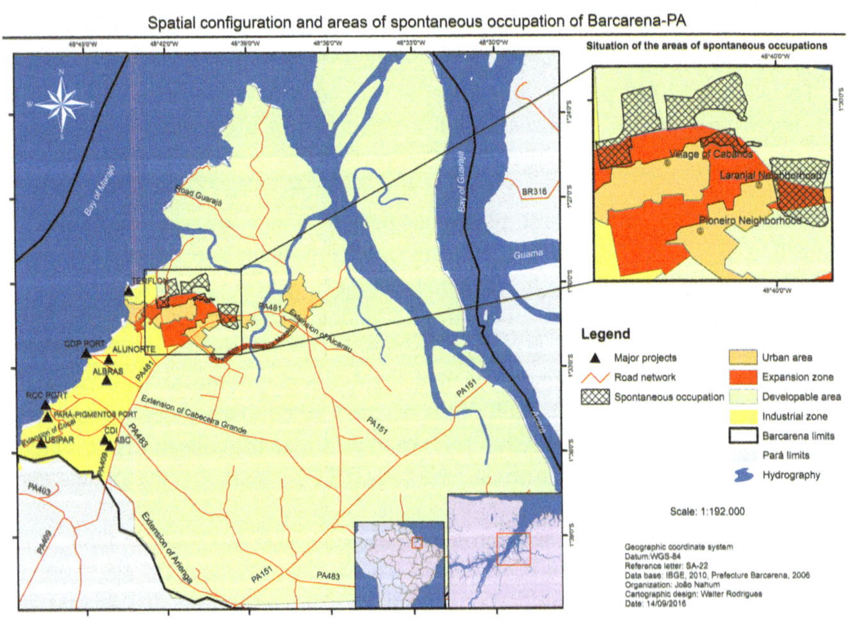

Map 2.1 Spatial configuration and areas of spontaneous occupation of Barcarena-PA

referred to as places of spontaneous occupation Bairro Novo, Novo Horizonte, Bairro Laranjal (Map 2.1). Are extend to the rural and coastal areas of the islands, where a resident population faces unemployment, poor housing and even food. As emphasized by a secretary of a trade union we interviewed:

> Because they have no jobs, some no longer have a home. Then you will see here in Barcarena the amount of invasions that occurred in the urban area is absurd, just here around Vila do Conde, around the urban center of the Village of Cabanos, which was a town planned for the middle class of employees of Albras, there you have the Cabano Garden, the Novo Horizonte, the Industrial district, Rio Paraiso, just around the urban core. You go to Vila do Conde, you have 9 de Abril, you have industrial Jucuraru, all are invasions, people who have no job, no home, have nothing. Sell the house because is unemployed, then invade the land in order to live, it's horrible here in Barcarena (Verbal information).[7]

Municipal management transforms the management bodies, particularly the social assistance office and the city council, in call centers and supply a range of requests. The "Duty Social Program," developed by the Secretariat of Social Action, held emergency consultations with various concessions, according to the need of the existing demand. Funeral assistance, birth certificate, routing duplicate ID card, allowance for immigrant families, basic food baskets, medicine deals with medical prescription among others, as well as several referrals of the poor to public and private entities.

The strategy orders, either by enabling welfare actions, whether by linking public interest with private reinforces the councilman's image as "a person who does or solves." And council members themselves take the "act of dispatch" as common share, as stated the former president of the Barcarena city council, Renato Ogawa:

> Because I think it's quite natural, because the alderman in our political scale, is the only politician that people have access to, he is the only politician that citizens here in the city, the citizens here the city can come and knock on the door of the house of and be heard. You find aldermen walking down the street, you find aldermen playing soccer, you will find him in the square. You do not find state representatives, or federal, you do not find here senators to talk to, or the president of the republic at your reach, you do not find the governor. Access to the mayor is already a narrower thing (Verbal information).[8]

The population believing that the councilor is "the person who does and decides," "executive agent" welcomes him expectantly, especially during rare itinerant sessions of the municipal legislative chamber. What would be a discussion forum, built on dialogue and aimed at better management of the territory, metamorphoses in an ombudsman in which people, individually or collectively, request from the Alderman what is needed. The population—for not knowing their rights, the culture of force "of always been that way" and disbelief in political as implementers of public policy—is going to an Alderman in a meeting to seek what they need

[7]An interview given in March 2005.

[8]An interview given in February 2005.

immediately. Thus, it is propagated a fiction democratic forum at the expense of implementation of public policies democratically constituted from the demands of the whole society—only possibility of building a fairer country, formation of strong citizens.

The rare itinerant sessions of the Council, which should be a moment of exercise of legislative transparency and exchange of information, boils down to two acts: the first presents the provision of services of all, when the Councilman explains what you are doing where, how and to whom; in the second, there are records of new requests for goods and services to individuals or to the community. According to council members, his political activities meet the demands of society, this does not require information, transparency and participation in the planning, they act as if society had no right to participate, as if this law does not exist.

The population resorts to the strategy orders and does not believe in the government as a common good manager, so his lack of interest in knowing and using their rights to participation in planning and access to information. Disbelief in politics was reiterated by a leader of the community center of São Francisco, in an interview granted.

> The community of disbelief is huge, the people they no longer believe in Councilman business, vote because you have to vote, vote because he is a citizen and has to vote, but not because he believes that so will bring some benefit, go claim, might bring to São Francisco because this here is years, years and years of either claim and ask, they say they will. This last week I've been with Paul Ronildo, who is the councilor, second term, who always said he would claim to make our place, to improve the water supply system, but I do not know why, but it does not come (Verbal information).[9]

The president of the city council, however, has (had) another version for disinterest. According to him, "it is created the culture that it's useless, you can only solves things if by their intermediation"—the councilors. The population is not interested in effective political participation, to know the role of the legislature, confusing it with the executive power. So the councilors do not direct actions to build a political understanding of their activity because they believe it is not the desire of the population.

2.1.2 The Political Coercion

At Barcarena, in general, the population is not involved in politics, even considering the political activity as a dishonest activity. They do not aspire to any office, position or function in the executive or legislative power. At most, they participate as members of the community centers, neighborhood associations, unions. The average citizen expects the ruling elite to continue dispatching and therefore criticized the former mayor, Laurival Magno Cunha, being president of the Consortium

[9]An interview given in March 2005.

for Social/Economic Intermunicipal Development (CODESEI), dispatched the seat of the body, in Belém, very away from the population that usually seeks.

The minority of the population participates in community centers, neighborhood associations, trade unions, and political parties and rarely appears a citizen willing to apply in municipal elections. The low participation of people in organizations representing their interests is a characteristic feature of brazilian political culture, as made clear by Santos Junior et al. (2004). In an interview we asked a union director: democratic management is a demand of the organized population or popular organizations aimed at solving specific problems of the category?

> This position of yours is right. But there's something else to make it happen that way. It is the political pressure. Because our bodies all have public officials. Those who are not a civil servant, have a brother, a family member who is a public official. So as you begin to show the mistakes of an administration or 'hit on her', or put the population in a conscious way is that you have to work, the person connected to your family begins to suffer pressure: "look, or your nephew or son or whoever stops do it here or you're going to the street" (Verbal information).[10]

A director of another union reported that kind of repressive and authoritarian practice of local governments:

> The government has a practice of trying to pursue, trying to frighten, intimidate. Ifthey don't get it, they set out to co-opt. The management that tried to co-opt the most was the current – Laurivalzinho – offering individual or group advantages, like, if you support me or if you do not challenge the administration we can [...] provide that, usually for clear leaders (Verbal information).[11]

Local governments, through the actions of conservative system, repress those who disagree with their interests or try to escape their command. On the other hand, still create the illusion that the democratic management is not a demand of society, "because no one spoke against the actions of the government." The population, organized community centers, residents' associations and trade unions, lives down and suppressed by conservative actions, it is paralyzed in the face of traditional political management and ultimately withstand the pressures of this. There are times the action of civil society is limited by this constraint, which is part of Barcarena political culture. In the words of another (former) union director:

> Just now, I spoke of the issue of culture. It is cultural the issue of non-participation in political life. This is not occasioned in the population. This, in a way, is reflected in the unions, in the general organizations. We host this thing over the years, these ideas, the ideological question, and consequently the practice (Verbal information).[12]

The political coercion favors the system of conservative actions preventing and fighting behaviors that threaten the power of the rulers. The population therefore can barely organize, trade unions and community associations most often provide

[10]An interview given in February 2005.

[11]An interview given in February 2005.

[12]An interview given in February 2005.

answers to everyday questions, ordinary, related to *their* category of work, *their* street, *their* neighborhood, *their* community and do not threaten to turn the base of the traditional political power. Reiterated one (former) union director: "here we operate more on what is put to us on a daily basis, or sometimes, the unexpected, unforeseen." They act motivated by specific interests, often individual or corporate.

When the shares have purposes and corporate or individual interests, society loses the prospect of performing as civil society, such as,

> a complex field of ideological struggles in which classes, interest groups and individuals, alone, seek to achieve hegemony, reform the state and influence its policies. This historical concept, rather than legal, civil society does not detract from the ethical content. It is in civil society and through the ethical and civilizing values are affirmed, in that groups that intend to carry these values (and possibly are) are part of it and on it seek to exert its influence (Pereira 1999, p. 100, author's translation).

Not realizing civil society has a vicious circle because, according to the union, the fight against conservative political power can only be effective if the neighborhood associations, community centers, trade unions, in short, the organized people plan their actions with collective perspective. According to another interviewee, one (former) union secretary:

> Trade unions and civil society are not in a condition to organize to discuss the public space, the improvement of public space as a whole, but specific actions. The problem appears, then you try to give an answer or say something about the problem. Civil society is organized more as a category, than as a citizen, and weighs the issue of corporatism. It is the category is not the citizen. Here we discussed a lot, our demands are many, but in the end what has been worth, which carries more weight, is the issue of wages. But we discuss the salary and not the budget, quality of life, public participation, administrative transparency (Verbal information).[13]

This, according to Santos (1993, p. 75), is a political action purely spacist, but not exactly spatial, for "acting on a fraction of the territory without the action to be designed in a comprehensive manner, can offer topical solutions and limited effectiveness in time, serving mainly the strengthening of structural data against which you imagined you were fighting."

Politically repressed, the civil society mobilizes to claim individual interests, does not respond to the processes of territorial reorganization, such as transition management through which passes the village of Cabanos, the implementation of public administration management, the transformations imposed now by the Fiscal Law responsibility. Monitoring, understanding, and intervening in rearrangements resulting from the implementation of new production systems engineering, transport, energy, and communication in Barcarena, as well as changes in the management of the demanded territory by establishing a more stringent regulatory system, such as the standards system of public finances to ensure responsible fiscal management, is the challenge for civil society.

[13]An interview given in February 2005.

According to an interview by a (former) union director:

> The speed in which large industrial projects come is very fast, they are installed and somehow run over the movement. The move would have to be planned. It has several aspects here, for example, one is the state of the infrastructure service for the region, the road handle. It seems not much, but it brings a reflection. From the road loop, we have a number of components here in the city that change from the point of view of violence, as in the aspect of economic issue. We have here situations, an example, companies that are, from the handle, feuding to get the best slice of the pie, road transport. So, what happens to the social movement? We get the trailer because it had not planned. And that, in a way, we already know why we did some discussions on the road handle ten years. At the time when it would be implemented here I had a first path, an initial project. On top of this project we discussed. But when we were discussing it, the ruling class was already discussing other projects. We could not keep up. In general, the movement is not following the changes (Verbal information).[14]

The population only watches the impact of territorial reordering processes because political inertia (of directors governing associations and unions) compromises the possibility of a broader and tougher mobilization. In fact, over the society suffers the strength of the political power of the rulers and below its own base and political organization follow at the same pace, driven by individualistic and self-interested view. Thus, the focus of fundamental concern of the unions are working conditions and salary. In trade unions, workers are treated as injured clients that have to be personally represented and defended by their class entity losingthe dimension of social class.

2.1.3 The Centralization and the Opacity

The centralization of power is the strategy used by the ruling elite to structure the municipal administration of Barcarena. It appears in the composition of the leadership of the management staff completed according to the interests and directions of the elected and families that formed the basis of political and economic support in the elections. In 1997–2000, management of Wandick Gutierrez, deputy mayor appointed the secretary of health; Secretary of culture and sport was given by João Carlos, vice mayor of Wandick Gutierrez in the first term; Secretary of Agriculture was indicated by the Councilman Fernando Amaral; the secretariat of works and transport was the choice of Mr. Pedro Smith Amaral, uncle of the Alderman Fernando Amaral and friend of Wandick Gutierrez, who was later replaced by one of the sons of Gutierrez, Carlos Alberto dos Anjos Vieira.

In the territory of Barcarena, the centralization of power of the ruling elites exemplifies a trait of Brazilian political patronage, as analyzed and conceptualized by Faoro (1991, p. 84), power centralization with patrimonial traits:

---

[14]An interview given in March 2005.

> Patrimonialism, basic political organization, closes on itself with the estate of markedly bureaucratic character. Bureaucracy is not in the modern sense, as rational rigging, but the charge of appropriation – the office charged with own power, combined with the prince, without the cancellation of the own sphere of competence.

Brito (2001, p. 33) reiterated that paternalism in Barcarena, where

> the political elite has exercised and still exercises a political patronage that reveals the predominance of relations of cronyism and patronage as mediators of political relations established in the municipality, which is administered in the context of a delegation of power, in which the political body represents a care institution, based on the principles of a private company.

The boundaries between public interests and private interests are dissolved. The retrofits in territory management are used by the local elite for their own private purposes. Modern standards of democratization of public management, based on the participation of the population and unrestricted access to information, set out in the Federal Constitution of 1988 and the constitutional reforms of the 1990s, are deployed in accordance with the system of conservative actions.

In Brazil, during the 1990s, constitutional reforms have reshaped the relationship between the union, the states, and municipalities, which are now responsible for providing social and guarantee of rights such as education, health and social care. The claims relating to the management autonomy and decentralization of government actions in favor of states and municipalities already included in the Federal Constitution of 1988. This was ensured through the tax powers of transfers and the increased share of these federal entities in the distribution of the collection of federal taxes.

> The transfer of tax powers implied the elimination of the EU's participation in the collection of the product of so-called flat tax, which was: (1) 40% in lubricants and liquid and gaseous fuels; (2) 40% in electricity and 100% in communications services; (3) 30% in transport; and (4) 10% in mineral; the triggering events such taxes were incorporated into the state tax calculation base called Tax Relating to Goods and Services – ICMS.
>
> They remained in the competence of the Union six taxes: Tax on Income and Earnings of any Nature - IR; Excise Tax - IPI; Tax on Credit, Exchange and Insurance or Involving Bonds and Securities - IOF; Tax on Rural Territorial Property - ITR; Import Tax - II; Export Tax - IE. The wealth tax, provided for in the Constitution was never instituted (Silva and Oliveira 2000, p. 14, author's translation).

The increased participation of states and municipalities in the collection of federal taxes caused a significant reduction in available revenue of the National Treasury. According to Silva and Oliveira (2000), the states have received, from 1993: 21.5% Income Tax (IR) and the Tax on Industrialized Products (IPI); and municipalities 22.5%. Moreover, the North, Northeast, and Midwest have received 3% of the collection of these two taxes, for use in funding that would go to the private sector. Finally, the states have received 10% of the collection of the IPI, in proportion to exports of manufactured products. Belonging also 30 and 70% to the states and municipalities, respectively, of the collection of the product Tax on Financial Operations (IOF) levied on gold, as defined by law as a financial asset or exchange instrument. Finally, it is intended to municipalities 50% of the collection of the ITR.

The sharing of responsibilities between the Union, states, and municipalities also provides instruments of direct participation for the population, both in the planning of administrative activities such as the monitoring and control of resources. Thus, the following laws were enacted:

(a) Federal Law 8.913 of July 12, 1994, which provides for the decentralization of school meals, ruling in its Articles 2 and 3:

> Art. 2. The funds will only be transferred to the states, the Federal District and the municipalities that have, in operation, School Feeding Councils, made up of representatives of local public administration, responsible for the area of education; of teachers; parents of students; and rural workers.
> Art. 3. It is for the School Feeding Council, among others, the supervision and control of application resources for school lunch, and the drafting of its bylaws.

(b) Federal Law 9.394 of December 20, 1996, which establishes the guidelines and bases of national education, introducing between the principles of education, democratic management, Article 3, VIII;
(c) Federal Law 9.424 of December 24, 1996, which provides for the operation of the Fund for Primary Education Development and Advancement of Teachers (FUNDEF), ruling in Articles 4 and 5:

> Art. 4. The monitoring and social control over the distribution, transfer and application of the Fund's resources will be exercised to the respective governments within the Union, States, Federal District and Municipalities, for councils to be established in every sphere within one hundred and eighty days from the effective date of this Act.
> Art. 5. The accounting records and management statements, monthly and updated for the funds transferred, or received to the account of the Fund referred to in art. 1, will be permanently available to the councils responsible for monitoring and supervision within the State, Federal District or Municipality, and federal, state and municipal internal and external control.

With regard to health services, with the introduction of the Unified Health System (SUS), the following laws were enacted:

(a) Federal Law 8.080 of September 19, 1990, which establishes the conditions for the promotion, protection, and recovery of health, the organization and operation of the corresponding services, ruling in its Article 7, Paragraph VIII, community participation as principle guiding the development of actions and services;
(b) Federal Law 8.142 of December 28, 1990, which provides for community participation in the management of the Unified Health System, by introducing the instance collegial Health Councils, giving them the following competence:

> Art 1. […]
> § 2. The Health Council in permanent and deliberative collegial body composed of government representatives, service providers, health professionals and users engaged in the formulation of strategies and control the implementation of health policy in the corresponding instance, including the economic and financial aspects, whose decisions will be approved by the chief power legally constituted in each sphere of government.

Article 4 also states that to receive the funds, municipalities, states and the Federal District will rely on the Health Councils. Finally, the organization of the social assistance system by Federal Law 8.742 of December 7, 1993, popular participation was introduced through representative organizations, in policy formulation and control of actions at all levels, by the composition of popular councils, which is one of the conditions for the receiving of funds.

As inferred by these standards, the social rights of municipalization are synonymous with transparent, participatory, efficient, and responsive to the demands of each place. Citizens, through sectoral policies management councils, have guaranteed the right to participate directly in the planning and control mechanisms and enforcement policies in education, health, social assistance.

The councils are composed of representatives of both the organized society and the government, resulting in a composition that ensures half of representation for civil society organizations and half to the representation of government agencies. Despite being linked to government agencies, boards have autonomy in its powers to define its rules and dynamics of operation, which should be established with the participation of all social segments represented within, including, therefore, the public power itself.

However, in the territory of Barcarena, boards in health, social care, and the rights area of children/adolescents were created and institutionalized, almost, to have secured the transfer of funds from other levels of government to the municipal level to implement social policies in the areas of health, social assistance, child, and adolescent.

The political coercion of the municipal government does not allow to form councils, for example, for agriculture and the environment. According to our interviewee,

> The dialogue with the government today there was no change. Since 1988 I work with the labor movement here, we have difficulties with the municipal government here in Barcarena. We fight already for both the implementation of the municipal council of agriculture, which until now did not work. The former mayor signed a law, the father of Laurival Cunha, when he left and started Gutierrez, the successor mayor, made another law that also went nowhere, then the movement has always stood apart because the agriculture department continues making government jobs without contact with organized civil entities. They created a way for some associations have people connected to them, the political group's own municipal government, and they have some support from the municipality […]
>
> So it works like this, the municipal council of environment, if you are asking the Secretary of Environment they'll say that it exists. For us it does not exist, we do not participate, was not even called a municipal conference of environment for us to discuss the municipal council. But this was not done with the civil authorities, this was done only with the secretariat and the legal advice of city, which is the prosecutor of the city, with saying them with the public ministry, which today we known whether the prosecutor participates or not. But we know that we, civil organizations, not participate. And they say there is advice (Verbal information).[15]

[15] An interview given in March 2005.

The action of the ruling elite transforms the few councils in formal structures, subject to the administrative routines of the municipal departments in order to meet the contract approval procedures and providing due account of agreements with state and federal programs. The advice may not submit different proposals made official by those departments. According to the same counselor, "if it is to create proposals there's no way, because the secretariat already has them." So the council works to approve what comes from above. "The counselor must sign all that the secretary is asking and that the mayor is asking, but it is only to approve the project, program, providing that office accounts or to hold meetings" (Verbal information).[16]

The councils, which have been proposed as standards that guarantee the administrative modernization of public administration, end up being well-founded fiction, fable, allegory without autonomy and power to influence the direction of municipal policies. The strategy of the rulers is to not have information available, or disclose to society how policies are managed. Therefore, without information and transparency, the population is unable to monitor and decide on the management of social policies.

Therefore, in the territory of Barcarena, management boards exist to play the system kept actions of the ruling elite. Set up originally as population participation instruments are used to strengthen the political centralization of municipal governments, as it allows the city to receive the funds linked to the sector of advice policy, without compromising the traditional way of managing them, without the demands of the population being considered, without the public becoming aware of the amount of money, where and how it is allocated.

This mechanism of the conservative elite allows the municipal management to adapt documents, Plans, Programs, Budget Guidelines Law and Budget Laws, forwarded to the institutions and organs of control and inspection, looking to fulfill what is expected of modern management, in harmony with the national law and attentive to municipal problems.

The Multi-Year Plan of 2002–2005 of the first management Laurival Magno Cunha (2001–2004) was used to show where and how municipal revenues would be applied. This Multi-Year Plan consisted of competence of the secretariats programs, such as Health for All; Infrastructure; Generation of Employment and Income; Education; Basic sanitation; Public security; Agricultural development; Social Action; Sports and Leisure Culture; Urbanism; Management; Participatory budgeting; School bag; Transport; Tourism and Environment. The goals of the programs showed that the rulers knew/know the problems faced by citizens and interest in the improvement of the local population's quality of life.

Among the goals of the program Health for All of the municipal health secretariat included the deployment of 04 ICU beds at the municipal hospital, the construction and equipping of health unit with 30 beds in the county seat. For the

[16]An interview given in March 2005.

period of 2002–2005 the first project included allocated R$ 600,000.00 and the second R$ 500,000.00. The city, however, until December 2004 had 81 hospital beds, the same number of beds in 1999, also it had 18 outpatient units registered in SIASUS, number lower than in 1999, when they were 21 units.[17]

Some goals of the Education Program looked like the targets of an infrastructure program secretariat:

(a) construction of a Technical School and University for the period of 2002–2005 and this project was allocated R$ 246,000.00;
(b) construction, implementation, and maintenance of a Higher Education Institute. For the period of 2002–2005 this project was allocated R$ 340,350.00;
(c) construction and maintenance of the building to the City Board of Education. For the period of 2002–2005 this project was allocated R$ 50,000.00;
(d) construction and maintenance of own building of the Municipal Education. For the period of 2002–2005 had been allocated R$ 395,820.00.

The Secretariat building goals of Education did not take place. According to IBGE[18] data on education, enrollment, teachers, and schools at Barcarena, by the year 2004 the municipal public did not count nor with a school of higher learning, nor with teachers and students. Nor had municipal technical school and higher education institute, nor teacher and student. As fanciful as the technical school building goals, the university and the institute of higher education is the construction and maintenance of the building of the Municipal Council of Education. This does not even exist in the form of municipal law,[19] moreover, not even the Department of City Education works in its own building, occupying the same building of the Legislative Chamber.

The Agricultural Development programs of the Municipal Agriculture is in the spatial ambiguity of the fishing pole deployment project in the Lake of the reservoir of Tucuruí Hydroelectric Plant, located in the municipality of Tucuruí, approximately 300 km from the city of Barcarena. Tucuruí and Barcarena are even municipalities belonging to the same macro-Para. However, we know that there is no fishing pole or on site support for Barcarena fishing colony, which allows us to interrogate finally: Where was spent the money provided for the project?

We could lengthen the description of the programs that make up the Multi-Year Plan of 2002–2005, but we think that the above illustrates the mode of action and objectives of the local ruling elites before the new federal standards of administrative management. They use the preparation of the Multi-Year Plan to present to the supervisory institutions where and how municipal revenues would be applied.

---

[17]Available in: http://www.sepof.pa.gov.br/estatistica/ESTATISTICAS_MUNICIPAIS/Mesorr_Metrop_Belem/Belem/Barcarena.pdf. Accessed on Dec. 01, 2005.

[18]Available in: http://www.ibge.gov.br/cidadesat/xtras/temas.php?nomemun=Barcarenacondmun=1501. Accessed on Dec. 01, 2005.

[19]Available in: http://www.ibge.gov.br/munic2001/index.htm. Accessed on June 22, 2004.

2.1.4 The City's Policy in the Works

For Santos (2005), geographic objects contain information and its distribution in the territory tells how this can be used. The author states that it is the stock system that gives life to objects, so an analysis of geographical space cannot take such systems separately. The management of the territory must be understood in the light of a set of reordering actions of object systems that gives us the territorial configuration of a place. For good or for evil, government policies make up a system of reordering actions of the territory of the whole society of a place.

In Barcarena, the reordering actions of the territory are carried out through an endless project of works, like the rulers have political action that responds to the spatial demands of the population. However, the landscape and the spatial configuration of the municipality denounce the use of the territory as a resource for maintaining its political dominance. There is never enough budget to complete the work, and it remains there, as propaganda of government actions; when completed are presented as a donation from the rulers of society. It must be said that the budget of these works is public money, therefore belonging to society itself, and not a gift from this mayor or that councilor.

But the opening is a solemn act, at which the ruler appropriates the work, naming it and presenting the barcarenense society. It reads in Para Daily newspaper of April 23, 2002:

> *Residents of Barcarena received yesterday a present during the commemoration of the 46 years of the mayor Laurival Magno Cunha* (PMDB). It was the beginning of construction of the retaining pier that will be part of a large tourist complex in the local edge. Former Senator Jader Barbalho, federal representatives Elcione Barbalho and Jose Priante, beyond councilman hélder barbalho, took part in the important event for that municipality.
>
> According to the secretary of Barcarena Works, Neville Marlos, the pier will be the fundamental work within a large redevelopment project in the city, which has a R$ 34 million budget, and will be completed in three years.
>
> The retaining pier, which has had its funds released (R$ 1.1 million), will be done in 391 days. The whole border will still have a panel overlooking the river, and a paved, food court, market (all informal trade will be relocated), and improvements on the beach of Prainha. The work will have 1300 m long. "There will be a waterway station with a mini-mall, to receive ships", detailed the secretary.
>
> [...]
>
> *The mayor Laurivalzinho, as he is known by the people, said on his birthday who received the gifts was the municipality's population.* The works should be a key milestone for the modernization and resumption of development of Barcarena [...] (JADER VISITA NOVAS OBRAS EM BARCARENA 2002, s/r, emphasis added).

The mayor uses the ceremony of inauguration to strengthen the fable of the need to modernize the city. His lacunar discourse "is based, precisely because it can not say until the end what you want to say" (Chauí 1990, p. 22, author's translation). The minister admits the city's stagnant situation and proclaims his birthday the "resumption of development of Barcarena." Such promises, as noted by Santos

(1993), waves, usually to individual or collective improvements and achievements not undertaken by previous governments or by other politicians, such that sound in the ears of voters as opposed attitude to what was before, and to define themselves as a form of progressivism.

This fabulous modernization is meant to be a territorial reordering. In the words of the secretary of Works "a major redevelopment project of the city." But what the secretary calls "city redevelopment" spatially covers a fraction of the place, the district headquarters of the municipality, and a location, the edge that matches the Cronge Avenida da Silveira. This, for approximately 1300 m, are distributed commercial houses, supermarkets, street vendors, pharmacies, gas station, bus station, city hall, legislative chamber, farm office, banks, municipal market, fair farmer, mother church square mother, dock, bars, cafeterias, sound system and advertising posts, taxi stand.

There, although the presence of financial and political institutions is remarkable, as well as traditional business houses, the primary task of the majority of those who work there is "to survive and ensure the daily family life and participate, as much as possible, in certain forms of consumption peculiar to modern lifestyle" (Santos 2005, p. 102). We have extracts from the center of the lower circuit of Barcarena economy, "essentially consists of forms of manufacturing that are 'not capital intensive', not modern services, usually supplied by the level of sales and retail and commerce on a small scale and not modern" (Ibid., p. 97).

In the avenue Cronge da Silveira is observed the recognition of the trade in the common spaces of the street, the sidewalk, the square and the pier, turning everything into a free show. To rearrange this dynamic, the city has "to start construction of the retaining pier, which will reinvent part of the city, setting it to large tourist complex. In the words of the then secretary of works, "the whole border will still have a panel overlooking the river, and a paved, food court, a market (all informal trade will be relocated), and improvements on the beach of Prainha."

The works of the pier are propagated as blessings of the municipal government, next to each work stand signs or billboards trumpeting made with its own resources, as if the city were the only resource provider, hiding such works consists of funded plans and programs with federal and state resources. The modernization work, conducted by the management of Laurival Magno Cunha, hid the authoritarian planning, because the population knew the plan works in the mayor's birthday, having information and possibilities to react. Thus, workers from the waterfront, merchants, bar owners, and street vendors are removed from their workplaces so that the retaining pier construction works are started.

The removals, according to urbanization secretariat, are justified by the need of "urbanization and sanitation of the free fair," to which the Multi-Year Plan of 2002–2005 allocated R$ 68,000.00 and for the design of "personification of the free fair stallholders" to which the same plan has allocated R$ 340,400.00. These works, report Para's Daily newspaper, would be delivered in 391 days after the anniversary of the mayor. The situation in which they were, in November 2005, made it look like other removals would cause a "cyclical" movement, because the site from where they are removed is reoccupied by a new worker.

In this conservative logic it is essential that urban works are in progress, although in many cases, they do not translate the demands of the communities, as its conclusion may well appear on the platforms of the candidates. *It is because devices such as that that Barcarena's territory organization is used as a political tool by the small circle of deciding agents of territorial management.* The corollary of this use has been maintaining a territory without citizens, unequal and divided. In it, the goods and services are distributed unequally in the territorial configuration and those that can travel to the district headquarters of the municipality and the village of Cabanos where such goods and services are available are able to enjoy. Social inequalities are, first, territorial inequalities "because they derive from the place where everyone is. Treatment can not be oblivious to the territorial realities. The citizen is an individual in a place" (Santos 1993, p. 123, author's translation).

References

Arendt H (2003) A Condição Humana, 10ª Edição. Forense Universitária, Rio de Janeiro

Brito SM (2001) Os Donos do Poder no Município de Barcarena: Passado e Presente. Monografia de Especialização do XV FIPAM. NAEA/UFPª

Chaui M (1990) Cultura e Democracia, 5ª Edição. Cortez Editora, São Paulo

Da Matta R (2001) O Que Faz do Brasil, Brasil? Rocco, Rio de Janeiro

Faoro R (1991) Os Donos do Poder. Formação do Patronato Político Brasileiro, 9ª Edição. Editora Globo, São Paulo

da Costa Gomes PC (2001) A condição urbana. Ensaios de geopolítica da cidade. Rio de Janeiro. Bertrand Brasil

JADER VISITA NOVAS OBRAS EM BARCARENA (2002) Ex-senador aceita convite de prefeito e vê de perto as transformações no município. Diário do Pará, Belém, 23. abr. s/n

Martins JS (1999) O Poder do Atraso. Ensaios de Sociologia da História Lenta, 2ª Edição. Editora HUCITEC, São Paulo

Milliband R (1999) Análise de Classes. In: Giddens A, Turner J. Teoria Social Hoje. Editora UNESP, São Paulo, pp 471–502

Oliveira F (2004) "O momento Lênin". O Pensamento nas Rupturas da Política. Relatório Final de Pesquisa Fapesp, Centro de Estudos dos Direitos da Cidadania, São Paulo. USP/FFLCH; Julho (mimeo)

Pereira LCB (1999) Sociedade Civil: sua democratização para a reforma do Estado. In: Pereira LCB, Wilheim J, Sola L (Orgs) Sociedade e Estado em Transformação. Editora Unesp; Brasília; ENAP, São Paulo, pp 67–116

Santos M (1993) O Espaço do Cidadão, 2ª Edição. Nobel, São Paulo

Santos M (2005) Da Totalidade ao Lugar. Edusp, São Paulo

Santos Junior OA, Ribeiro LCQ, Azevedo S (Orgs) (2004) Governança Democrática e Poder Local. A Experiência dos Conselhos Municipais no Brasil. Revan, Fase, Rio de Janeiro

Silva PF, Oliveira JC (2000) Reformas das Instituições Fiscais: Reflexões sobre o caso do Brasil. Disponível em: http://www.eclac.cl/publicaciones/DesarrolloEconomico/6/lcl1366/lcl/366e.pdf. Acesso em: 8 June 2002

Chapter 3
The Management of Laurival Magno Cunha and the Implementation of the Fiscal Responsibility Law

Abstract In the third chapter I examine the reproduction of conservative actions system in each process of modernization of the territory of Barcarena, focusing on the process at the time of implementation of the Fiscal Responsibility Law in management (2001–2004) of Laurival Magno Cunha. This government did not use the law as a fiscal adjustment instrument, as conceived by the global reason. Suited to his interests the planning, public participation, and transparency established by Law, to show up modern and hide his conservative style, anti-democratic, and elitist land management.

Keywords Reproduction of conservative actions · Fiscal responsibility law · Public participation

The previous chapter showed how the elite uses territory management according to their particular interests. The practice of local political and economical elites unifies the class interests, focusing on the refusal of modernizations involving decentralization of power and democratization of public management, in short any act that threatens their political and economic power. In the face of modernization in the management of public affairs, sectoral social policies management councils are deployed in Barcarena, still according to the old system of conservative actions to ensure the continuity of the hegemonic power of the elite management and territorial control—the policy orders, the welfare actions, the detention of civil society, and the use of municipal revenues for patronage purposes.

In addition to these old practices other strategies are used by municipal governments. Now, they are the new rules (LRF) which serve as a resource to propagandize the most modern and democratic management of the political elite: Plans, Programs, Budget Guidelines Law, and Annual Budget Laws sent to federal oversight monitoring bodies and institutions These are a fable of what a modern administrative structure should be.

In this chapter I argue that a system of conservative actions reproduces itself at every moment of the process of modernizing the territory of Barcarena. I show how the Fiscal Responsibility Law has been interpreted in the 2001–2004 term of

J.S. Nahum, *Modernization and Political Actions in the Brazilian Amazon*,
SpringerBriefs in Latin American Studies, DOI 10.1007/978-3-319-58030-2_3

Laurival Magno Cunha. This government did not use the law as an instrument of control and fiscal adjustment. Cunha's Government suited this planning of public participation and transparency established by this law to their particular economical and political interests to seem modern and hide his conservative, anti-democratic, and elitist style of territorial management.

3.1 The Fiscal Responsibility Law: A State's Instrument of Reform

The Fiscal Responsibility Law, Complementary Law 101 of 04 May 2000, is a set of modern rules designed to accomplish certain purposes determined by the global order created by the world.

> This is the world of the universal market and worldwide governments. The IMF, World Bank, GATT, international organizations, the World Universities, Foundations, which use money to heavily stimulate research, are part of the world government that intends to deploy, giving grounds for perverse globalization and to the attacks made today that, in practice and ideology, are directed towards the territorial State (Santos 2005, p. 142, author's translation).

This global order requires the governments' national standards and guidelines for economic policy which should adapt, on the risk of losing the possibility to participate in the modernization process, understood as the most competitive way of global market participation. In the central capitalist countries since the 1960s, the capital of reproduction has required fiscal adjustment from the states. What meant the reduction of welfare state policies that for Giddens (1999, p. 16–7) in social democracy is characterized by diffuse involvement of the state in social and economic life; domination of the civil society by the state; collectivism; Keynesian demand management, added to the corporatism; restricted roles for markets; full employment; strong egalitarianism, comprehensive welfare state, protecting citizens; linear modernization; internationalism; etc.

For the reformers, social welfare policies (welfare state) were responsible for the public debt of states and the heavy burden of taxes that companies must pay, which caused slow growth of productive capital. In this scenario, the fiscal adjustment was defended by the institutions representing the capital and presented as standard to all territories must adapt, reducing the state to its basic functions. In summary,

> the main, if not the only, goal of the federal government with the approval of this tax law is to raise funds to pay the Brazilian external debt. To meet the requirements arising from the IMF, the World Bank and the United States, the Federal Government introduced the Complementary Bill 18/99—Current Complementary Law 101/2000—including it in its structural character of reform objectives set in the ongoing program of fiscal and macroeconomic stability (Castro 2001, p. 17, author's translation).

In Western European countries, the state reforms were preceded by a long period in which the social welfare systems were in force with training proposals of a strong civil society, with expanded and secured civil rights, respecting political rights and founded on freedom. In Brazil, by contrast, a long period of military dictatorship has undercut political freedom and was responsible for the dismantling of attempts to build citizen participation and social welfare policies, leading to the current imposed on the state reforms. The brazilian society does not even have its guaranteed minimum rights and the rule has proved exemplary in closely following the adjustment proposals of public accounts, even though we know that in most cases it has directed the public fund to meet the interests and demands of international capital. Since the early 1990s proposals for liberal reforms were regarded as essential to the viability of the country. In the name of the global imperative of the country participating in the process of modernization, reforms settled and quickly much of the public property passes into the hands of private companies.

One of the state's reform objectives of the proposal is to ensure continuity and increase the public funding allocated to the reproduction of capital. One way to accomplish this is the regulation of state assets allocated to social spending and/or that make up the indirect wage. Fulfilling up reforms, as is being proposed, we involve in social relations of production, putting them on a level surpassed long ago by the welfare state when it created conditions for the commodification of the workforce. The value of this was not only determined by market dynamics, but indirect wages were added in the form of public fund allocation ensuring education, health, welfare, unemployment insurance, and leisure, among other social rights. It is on this front that the attacks of the reformers focus when they classify the social policies as costs, as obstacles to attract investment. Therefore, in the logic of this new legal and regulatory system of government actions, this is the system that "must be reformed urgently because the economic gains of investors can not wait" (Bourdieu 1998, p. 64, author's translation).

Guided by economic determinants of the world market (financial market, stock exchanges, commodities and futures, monetary, fiscal and commercial central capitalist countries, etc.), the proposed reforms are exogenous guidelines to Brazil and do not take into account the specificities of the formation of national capitalism here that has, obediently, through the federal government, promoted the undisputed reduction of the public; all accompanied by the fallacious discourse of democracy of the market, decentralization, autonomy, transparency, accountability, and control. Fiscal adjustment is the goal, and it is in its name that "the imperfect citizen is transformed into the perfect consumer" (Santos 1993).

The Master Plan of the State Apparatus Reform, produced and published in 1995 by the Ministry of Administration and State Reform (MARE), proposes an important State Reform in Brazil.[1] A program ratifies an agenda of common

[1]This is not the first reform program. de Oliveira and Silva (2000), in Reforms in the Fiscal Institutions: Reflections on the Case of Brazil, propose a periodization of reforms in tax institutions. The Master Plan of the State Apparatus Reform there is a summary of other attempts at reform. We highlight the reform from the 1990s, the transnational character that it takes, and that it is contained in the premises for the building of the LRF and the construction of a new monetarist management of the territory.

transformations between Brazil, European countries, and the United States. According to this plan, in Brazil, "because the development model that previous governments have adopted the state deviated from its basic functions" (Brasil 1995, p. 3) and expanded its presence in the market leading to increasing distortions in this, now live with unsustainable artificiality in the 1990s. These deviations would justify reforms and add that they are necessary in "a result of the unprecedented bureaucratic setbacks of the 1988 Constitution" (Ibid, p. 18) such as earmarking revenues, social rights, social security, and labor legislation especially for civil servants.

In this sense, reforming the state apparatus:

> Means ensuring that appliance with greater governance, i.e., greater capacity to govern, the greater condition to implement laws and public policies. It means making more efficient the exclusive state activities through the transformation of the authorities into "autonomous agencies" and making much more efficient the competitive social services to turn them into non-state organizations of a special kind: "social organizations" (Ibid, p. 42).

The guiding rationale of the proposal is the managerial public administration. In the strategic core of the effectiveness, it is more important than efficiency, but in the sector of exclusive state activities, non-exclusive services and the production of goods and services, the efficiency criteria are fundamental. Although the efficiency criteria give priority to serving the citizens, its citizen's design is based on the idea of citizen–customer, consumer, and user (Table 3.1).

The strategic core of the proposal corresponds to the actions of government where strategic decisions are taken, defines the policies, and ensures compliance with these. The reform proposal covers the legislative and judicial powers, the prosecutor and the executive branch, the president of the republic, the ministers and their staff, and direct reports responsible for planning and formulation of public policies. In this core, the property must be necessarily related to the state and in order for its functions to be performed they need to rely on the support of activities or ancillary services: cleaning, security, transportation, computer technical services, and data processing, among others. According to "the years of state reform of logic 1990, these services should in principle be outsourced, i.e., should be subjected to public bidding and contracted through third parties" (Pereira 1997, p. 29).

The exclusive activities sector includes actions that only the state can perform among which the act to regulate, supervise, and encourage, for example, the collection and control of taxes, police, the basic social security, the enforcement of

Table 3.1 Institutions resulting from the reform of the state

	Unique activities of the state	Social and scientific services	Goods and services for production market
Main activities (Core)	State As personal	Non-state public entities	Privatized companies
Auxiliary activities	Outsourcing companies	Outsourcing companies	Outsourcing companies

Source Pereira (1997), p. 30

health standards, transit service, health services, control of the environment, the allowance to basic education, and the passport issuing service. The important thing is the quality and cost of services provided to citizens and the principle that should prevail is the search for a great relationship between quality and cost of services made available to the public. Therefore, the administration must necessarily also be the management. The same is true of state-owned companies that must comply with the management principles of administration.

The non-exclusive services sector corresponds to those provided by both the state and public non-state and private. These services involve sectors such as education and health, or those sectors that have significant external economies, as they produce gains that may not be appropriate for such services through the market—such as the foundations or NGOs, universities, hospitals, research centers, and museums. In this the ideal property sector is the public non-state. This is not state property because their state power is not exercised, neither it is private property because it is a type of subsidized service, often by the state. This non-state public property aims to make social control easy and direct, through participation on the boards of the various sectors involved, while favoring the partnership between society and state. A "situation where the market is clearly unable to perform a series of tasks, but the state also does not show sufficiently flexible and efficient to perform them, opens space for non-state organizations" (Pereira 1997, p. 28). This process comprises the transfer of a number of activities in the social responsibilities of the state to non-state organizations, which would no longer be exclusive to the state, which is called a process of "publicness". Through publicity, such organizations enjoy a much greater administrative autonomy than that possible within the state apparatus and all taxes that private institutions are subjected to.

All the reform model is based on the concept of the state as manager/ administrator, where basic social functions are kept to a minimum by transferring to the public non-state organizations to control key activities to society. The managerial perspective of public administration is in harmony with the idea of fiscal adjustment and have the principle that "(...) it is necessary for the state to reorganize and for that you need to adopt management criteria able to reduce costs, to seek greater coordination with society, democratically define and collect results" (Cardoso 2001, p. 16). Reduce costs, dialogue with society, set priorities democratically, and demand results have as instruments planning, transparency, control, and accountability.

3.2 The Fiscal Responsibility Law in Barcarena: Modernization and Conservation

The Fiscal Responsibility Law (LRF) is a state reform instrument and comprises a set of rules that guide the actions in the field of economic policy. The instruments of planning and budget execution as Multi-Year Plan, the Budget Guidelines Law, and the Annual Budget Law predate the LRF and have their general definitions set out in Article 165 of the Federal Constitution. These tools are becoming more important due to the indebtedness of the federal entities and the collection of society, tired of seeing their equity disappear without being reversed in its favor the necessary provision of services such as health, safety, and education (Castro 2001).

The Fiscal Responsibility Law is primarily a fiscal adjustment instrument of public spending. Regulating how, how much, and where governments at any municipal scale, state or federal, should apply the resources, taking away from the places the political autonomy to decide. Paradoxically, the proposed decentralization centralizes policy decisions through a normative control. The law carries the global order that seeks to impose everywhere a single rationality, the global market, and a unique way of solidarity, organizational, imposing on the places the "compulsory need of performing common tasks, even if the project is not common" (Santos 2005, p. 158).

The process of regulation of municipal governments through the implementation of the Fiscal Responsibility Law throughout the country has, however, taken any other contours. In my study of a place, Barcarena, I analyze how a conservative actions system reproduces there, before this process of transformation of administrative management of municipal governments. Planning is one of the pillars of the Fiscal Responsibility Law. Paragraph 1 of Chapter I of this law warns that responsible fiscal management presupposes the action planned and transparent, in which way are prevented risks and corrected deviations that may affect the balance of accounts and obeying limits and conditions regarding the resignation of revenue generating staff costs, social security, and others consolidated and securities debt, credit operations, including by advance of revenues, guarantees, and registration in remains to be paid. Chapter II relates to planning and new rules for the preparation's the Budget Guidelines Law (LDO) and Annual Budget Law (LOA). In structuring logic of the Fiscal Responsibility Law that should be used by governments as a strategy to reduce costs and make more efficient the administration of state services, and as an instrument of protection of public property against the interests of particular groups or open corruption, in short to address the fiscal crisis.

In Barcarena these standards of planning and public administration, far from used as fiscal adjustment tool and generating a primary surplus, are suitable to reproduce the interests of the traditional elite, now with a new discourse, coated, and modern look. One of these adaptations is given in the formal field of administration, when documents are sent, Plans, Programs, Budget Guidelines Law and Budget Law institutions and control and supervisory bodies, in which it does seem that the municipal administrative structure is in harmony with national law and is not indifferent to the problems of citizens. But this normative harmony as the discourse of modernization is a fable.

The planning documents issued to the courts of accounts, municipalities, and the state, denote a perfect match between the operation of the municipal administration and establishing the Fiscal Responsibility Law. According to the logic of this law, for example, the Multi-Year Plan of 2002–2005 is to continue the goals of the Multi-Year Plan of the previous management. We read in the preamble of the Multi-Year Plan 2002–2005:

> We observed how the effectiveness of the PPA 1998–2001 was able to change the culture of public administration, placing the city of Barcarena in higher level of development, especially with regard to the quality of the urban environment, reaching the entire municipal territory, associated to the significant amount of resources that have been devoted to actions in the field of social development, as can be seen (Barcarena 2001a, p. 9).

Why Multi-Year Plan 1998–2001 was able to change the culture of the municipal administration? How? What facts and figures show that? What levels of "high growth" this plan provided? As the Plan has contributed to improving the quality of the urban environment? What improvements? As they reached the entire municipality? The Multi-Year-Plan 1998–2001 should answer such questions. However, everything seems to fiction because insistently we require and we have not received any document proving the existence of the Plan 1998–2001. No municipal agency knew of its existence, nor Barcarena Prefecture, the Court of the State of Pará accounts or the Court of the State of Pará municipalities, even the councilors know the existence of the Multi-Year Plan 1998–2001, nor the legal advice of the Legislative Chamber of Barcarena know why such a document was cited in the Multi-Year Plan 2002–2005.

It appears that the ruling elite suitable planning established by the Fiscal Responsibility Law to the conservative actions system has invented the Multi-Year Plan 1998–2001 to show that the Multi-Year Plan 2002–2005 continues; confirm the existence of the practice of administrative planning and, especially, demonstrate that the administration is in accordance with the federal law. The Multi-Year Plan (1998–2001), even though fiction, serves to emphasize that the ruling elite adapts to modern standards of management and municipal planning.

In Laurival Magno Cunha management planning constituted not an instrument to establish the fiscal adjustment as objective the Fiscal Responsibility Law. According to the report of the then municipal secretary of finance management, Mr. Paulo César[2], "the first year was a little more complicated because the team that came did not find anything here. To get an idea you had no information about anything. To computer hard drive the previous government took, which made it very difficult. But neither was sorely missed, because we knew we had this plan, this follow-up work."

Thus, reports the same former secretary:

> You know that this issue of the Fiscal Responsibility Law is still very early, very recent. There was experience. What we did here was to make the Multiyear Plan (PPA) and other laws such as LDO, which is the Law of Budget Guidelines and the budget itself. We sought mainly down to the lowest level of administration, i.e., discuss with the secretaries and the secretary teams […] So, we sought from the proposal of the mayor government, we sit with the directors and secretaries and try to make that possible in terms of resources, estimating costs and trying to allocate resources. From there we had the LDO, it was immediately for the following year and the PPA that was already for four years.
>
> So that was the first time. And the second, which has involved the LDO and annual budgets, we have had the opportunity to from the experiences of the first year and

[2]An interview given in March 2005.

> everything else, try, shall we say, to inform the public. We try, through the action of departments with local communities, extract, especially in the first year, what were their needs: look there in the Piedade location is the road, so we put the road, is a health center, is a school.
>
> So from the first year experience we started to focus on this and then we started doing. Because only now at the end of his mandate is that it created by law to planning department that will implement, try to put into practice now, which will allow you to have conditions to do that work permanently, regardless of who is the mayor. Because what we observe here is the following, as some thinkers says, that process runs and runs and. This happens a lot in public administration, so to avoid this kind of problem, to depersonalize, it will give a framework for the planning department to allow it to have automatically conditions develop. Then you will be able to do, for example, those popular consultations, those mechanisms that you can listen directly to people without intermediaries (Verbal information).[3]

In Barcarena, the unavoidable act of planning the public administration has been touted by Laurival Magno Cunha to distinguish his administration from that of his predecessor, Wandick Gutierrez, the program that the government of Laurival Magno Cunha supported the drafting of the Multi-Year Plan 2002–2005. This program is registered in the Registry Office of the 1st Extra Judiciary Barcarena County. With the registration stamp reads, "These commitments can not be changed because they are registered in the office of the 1st Letter Extra Judiciary of Barcarena's County. This shows the seriousness of our government program and to serve as a collection instrument next to us, in the four years we will be managing the Barcarena."

Transparency, publicity, and accountability are the foundations of this political act you want to make the government program into law. The program, with campaign tool, was released and discussed with civil society, with the unions, community centers, neighborhood associations, and student organizations. The contents of this document have the promise recorded function, serving as a collection instrument, as if the population of this only willing to defend their rights.

Although the government program to take law creates an air of confusion and the seriousness of their makers, in fact, is the Multi-Year Plan that has the form and function of law, and instrument that should be used by the population to collect government unfulfilled promises. The multiannual plan is a policy management tool, more serious and different from the established government program intentions and dosed claims by the political elite.

According to the rationality of the Fiscal Responsibility Law, the plan, as technical and political planning tool, should be developed through public hearings, with popular participation, which would be listed goals for management. In this rite the preparation of planning would promote the participation of society in public administration. Democratic assumptions of the planning and management of cities emphasize popular participation enshrined in the Federal Constitution of 1988. The public involvement in the planning would be guaranteed by municipal leaders

[3]An interview given in March 2005.

through hearing and public consultation, referendum, plebiscite, representation in court, and collegiate bodies, among other forms of strengthening citizenship. The viability of public participation requires public access to information necessary for the effective exercise of their right.

The logic of the Fiscal Responsibility Law institutionalizes participatory budgeting. A strategy that goes beyond the act of communicating to the public about public spending is to plan and monitor government actions. The population, from the moment the mayor is elected, should participate and decide on the construction of the multiannual plan, which will extend by 3 years the management of the developed and the first year of the next administration. After discussed and organized, the plan will be submitted for consideration and approval by the legislature, taking then the form of Municipal Law establishing the guidelines, objectives, and management targets for capital expenditures and for those relating to programs continuing duration.

Participatory budgeting is a strategy to plan the use of public resources, punctuating where, how, and when revenues are allocated. The budget has a public character not to be a law, but to be prepared and approved in a public space, through discussions and amendments made by council members in the sessions of the House. The 1988 Constitution defines three integrated tools for the preparation of the budget, aimed at planning the actions of government. They are as follows:

1. Multi-Year Plan (PPA): predicts expenditure programs, works and services from, lasting over a year. In the first year of government, the mayor must propose guidelines, goals, and objectives that, after approval, shall be effective in the next 3 years of his administration and the first year of the next administration. PPA leaving the targets for each year of management.
2. Budget Guidelines Law (LDO): from the PPA, defines the goals and priorities for the coming year. The LDO also defines the rules on changes in tax laws, finance, and personnel, and establishes guidelines for how to prepare the annual budget.
3. Annual Budget Law (LOA): is the budget itself. It contains programs, projects, and activities that address the goals and priorities established in the LDO and indicate the resources needed for their fulfillment. Thus, defines the sources of revenue and authorizes the expenditure, expressed in figures, detailing them by government agency and function, may authorize the provision of additional credit or making loans by the mayor without prior authorization of the Board.

The participation of the population must be guaranteed in three stages. The preparation of the LOA, however, is the most appropriate step to initiate public involvement in the budget preparation process. The first development of participatory budgeting stage has internal character of negotiation between the various departments, particularly the Social Policy and Planning. It is essential to evaluate the potential of collection and municipal spending. Then, the municipal government must define what your short, medium, and long term is, to submit proposals to the population.

The following should be developed the foundations of the process of participation. Each municipality develops an appropriate methodology for your situation, looking to include

(a) decentralization of discussions with the population through municipal division in localities;
(b) the development of service criteria of the demands of the localities;
(c) the format of the population participation criteria;
(d) the definition of instances of participation and their specific expertise.

When resources are scarce, it is a moment of intense negotiation of the various interests to be served. A good tool to improve the quality of this dialogue is to promote visits to locations so the representatives elected by the people to broaden the level of knowledge of each site and his relationship with the municipality, generating co-responsibility in deciding the allocations of resources available.

Even with regard to the preparation of the budget, it is necessary to establish and fine-tune the relationship between instances of popular participation and government decisions, and between the participatory budget process and the city council, as the participatory budget process does not end with its preparation, which is paramount in the creation of mechanisms enabling the population monitoring of budget execution and oversight of government spending.

However, the ruling elite Barcarena fits all these moments of preparation of the Multi-Year Plan of 2002–2005 process, so it does not threaten their political hegemony and especially the direction of their private interests. Avoiding being accused of authoritarian and centralizing, the management of Laurival Magno Cunha (2001–2004) said there is popular participation in development planning. Even said that "initially were created programs of interest from various areas of government as subsidies to the elaboration of a journey of seminars where these programs were discussed and improved" (Barcarena 2001a, p. 9, author's translation). In the face of state control institutions, the government conceives itself as democratic and modern, including having his administration supported in the Participatory Budget Program that "communities, organized and democratic, decide the priority of works within budget will into account the number of inhabitants and the degree of need for each neighborhood, and the people supervise the implementation of public money" (Ibid, p. 43, author's translation).

The Multi-Year Plan 2002–2005 should detail the Participatory Budget, and punctuating would deploy where, when, how, and how much would be allocated to human and material resources, detailing that does not exist, or the Law of Budgetary Guidelines, nor the Budgetary Law Annual. Participatory Budgeting program appears as a commitment to campaign and not as continuous duration of program management, so that the text of the Multi-Year Plan 2002–2005 regarding the participatory budget is identical to the government program.

In fact, in Laurival Magno Cunha management (2001–2004) the participatory budget did not constitute a continuing program duration, or better, he was non-existent, as confirmed by the then union director and then president of a community center:

> In the government program, the document that was distributed and they call themselves the government program, still in the first management Laurivalzinho (2001–2004), there was the issue of participatory budgeting. That was the word at meetings, he would rule with popular participation, and it really did not happen in no time (Verbal information).[4]

> […] Participatory budgeting may exist between them, but the practice in the communities there. The communities do not have access to it, council members are aware of it, but this group of councilors is closed with the mayor to because if I want to claim any critical situation of the neighborhood I can not. For me to do this, if I go to the camera do the claims I will take a number of problems that the city council does not want to hear […] (Verbal information).[5]

In the preparation of the Multi-Year Plan 2002–2005 democratic forums were not used, roving camera, popular tribune, public hearings, and public consultation. The absence of these forums is insistently reiterated by leaders of community centers and unions. By asking a then union leader, was even with organized society, there is no dialogue with city council about what is the Multi-Year Plan, what are your goals? The response of our interviewee,

> No, because we only pull the discussion in the chamber when the bodies are positioned. Now the other management at the time of Gutierrez had a space of 15 min to pronounce the camera on certain issues that the unions wanted to report on agriculture, the agrarian question, or any other matter. This space was taken by the current council members who were elected in the current administration.
>
> We asked: But has the popular tribune…
>
> Was taken, this popular platform does not exist today in the town hall of Barcarena. You can only rule there, a leading civil society, a councilor cede his place, his space, his time. If Councilman find that you will speak perhaps the administration or any person or that will not hit right with the plans of the camera and public administration you do not speak, neither will give word.
>
> We asked: What about the traveling camera?
>
> That does not exist. This worked in government Gutierrez twice, one was in Curupere, where the community was threatened to get out of there, just out, threatened by Kaolin companies, which were being implemented. Pressured by the Industrial Development Company of Para (CDI-PA). Then there was made by the union, we request a special session that was done there. The other traveling there in Guajarino in Trambioca Island, where he was discussing who might come to settle in Trambioca Island a pipeline project that came from the Amazon in here, that was all the years of Gutierrez's management. From there to this government Laurival, this camera was not any traveling done (Verbal information).[6]

The responses of the former director of the Community Center of Laranjal Neighborhood and the then leader of the Pioneer Neighborhood Community Center were also interviewed. How could I raise these conversations with the people in times of structuring the Multi-Year Plan (2002–2005), the government of Laurival Magno Cunha (2001–2004) dismissed the dialogue with neighborhood

[4]An interview given in February 2005.

[5]An interview given in March 2005.

[6]An interview given in February 2005.

associations, community centers, and trade unions, i.e., in the words of a leader of the community center of Pioneiro town, "the participation of the popular masses in the municipality of the project, I tell you this because I know the city, I walked in election campaigns in any municipality does not exist in any locality proposal or craft by the government for the population to take part of the project." Neither the municipal councils, which are channels of participation and representation of the population in specific public policy management, were heard. But did little good, since they have no autonomy and act as breeding instances of elite policies. In the words of a municipal official in an interview in March 2005,

> Today the Mayor was keen to win and put all the people on the boards, I tell people you trust. Even today, during the election of the Child Protection Council, was a similar campaign with the alderman, like Alderman, with possible financing for the people that the mayor hinted could win. And he succeeded. With the union, he here and there, he put a job, a contract for wife, for brother, to immobilize the reaction. The same happens with some neighborhood associations, community centers. He has created ways to co-opt the leaders and silence reactions. This process of co-optation we didn't did.[7]

planning targets are not set from the popular consultation. Asked the then councilor Alirio[8] who prepared the Multi-Year Plan (2002–2005), how were chosen and established the goals of this plan?

> The PPA is drawn from the needs and specificities of each office. I went from 2001–2003 agriculture secretary had to move away to be candidate for councilor. We did a whole survey in rural communities, socioeconomic and cultural survey, which we call until then study of reality. Based on this study of the reality of both the rural and the complementation of agriculture that would fairs and markets and everything else we begin to trace the PPA divided into goals. […] So what we did, study of reality then we detected what, various community issues, both in social issues, health, education, and in the production issue that is our case […] So the community is key to show you what you really need to work and that's what we did […] So, you see, based on these surveys we made this program and shared the steps, based on a legal apparatus.
>
> I ask: How society participate in the elaboration of this plan?
>
> It is through the study of reality, which are questionnaires that we formed, around 70 questions about where you raised the whole population, where you wondered if it was the owner of the land, how many years lived longer there than he lived, the he extracted from the earth, which he planted, how many people in the family, how many married, how many children, how many are in school, it has suffered some disease, what kind of power they use most, are questions that can be said to be really everyday, but they give then an X-ray, the profile of the community, both in the matter of health, as in education and generation of employment and income. Then the society is important because it gives you all the information.

It appears that the system of conservative actions used by the Laurival Magno Cunha management (2001–2004) structured a traditional multiannual plan, because it ignored all the actors of the social process, a government that had as its starting

---

[7] An interview given in March 2005.

[8] An interview given in March 2005.

point the government program and disregarded the demands of the population. Besides, it was technocratic, as used diagnostics to explain reality, judging objective, scientific, and rigorous. Then required the company guidelines and elaborate orders based on statistical surveys and according to the priority of each office.

This traditional style of preparation of government plans an old-style rule whose legitimacy deficit deepens as the decision-making process involving a small circle of cronies, far removed from the interests of the entire population. This was confirmed in the words of former Secretary of 2001–2004 management of finance, Paulo Cesar: "I can not tell you that we can make the people participate in the development. I was being flippant if I said that. There was no participation of the population" (Verbal information).[9]

We could consider that governments have undertaken efforts to implement the plan as stated in the LRF, with transparency and participation, stimulating participation strategies were used, that the media were triggered, in short everything he could to reach the elite ruler was used for the population to participate. But the population, little interest in political life of the city, did not participate and then prepared the Multi-Year Plan (2002–2005), even without the participation of the population.

The finance secretary frankly dismissed the functioning of the conservative actions system in municipal management. Incidentally, this same lacunar discourse strategy was used by former councilman and mayor Renato Ogawa, when asked, as they were chosen and established the actions and goals of the Multi-Year Plan 2002–2005? He answers "Unfortunately we have here little participation of civil society" (Verbal information).[10]

According to the former councilor and former finance secretary, the population is solely responsible for not participating in the planning activity. As if the lack of participation was an individual and free choice of coercions policies; as the development of public policies was the main activity of the rulers; as the mayor, secretaries, and councilors make available or legitimize discursive arenas, roving camera, popular tribune, public hearings, public consultation, in short, democratic forums in which the population, regardless of ideological conviction, political and partisan could express their opinions without fear of retaliation on the territorial management; as if the people living in a territory which has autonomous media and the ruling elite make available information on the budgetary situation of the municipality, where the budget is being used and how to monitor the implementation.

Far from being public domain, information about government actions are not accessible to the ordinary citizen. For this, the source of information is said to be the news that publishes events behind the scenes of the municipal administrative structure, whose source is often an officer of the legislature or the executive. The *news*, the subject of everyday conversation, notify the company before the media on

[9]An interview given in March 2005.

[10]An interview given in March 2005.

likely candidates for executive and legislative, political agendas, political visits to localities, establishment of political party agreements, social action programs, agreements between business elite and ruling, and arrival of new ventures, but its range is absolutely bound to report the fiscal situation of the municipality, especially its income and expenses.

The news is a very different expedient of establishing the Fiscal Responsibility Law, which assumes the existence of transparency and civil society participation, in the sense that the plans, the Budget Guidelines Law and budgets, accountability and its previous opinion of the Court of Auditors, the Annexes of Fiscal Targets and Fiscal Risks and Budget Execution Report and Fiscal Management, together with its simplified versions should be available for consultation and examination of all citizens.

However, all this is far from the political everyday territorial management Barcarena, council members (of the time) explained that the lack of public participation in the political activity is an individual attitude, barcarenense citizen does not like to get involved in the political life the city, and it is not his interest.

According to the councilor (at the time) Paulo Cesar,

> […] a minority of the population has a desire to participate. She even sometimes charges the government and councilors. *But the vast majority do not, however, if you call it participates.*
>
> I ask: The government has encouraged the company to participate in political life?
>
> As I told you there was so much of the executive, because it lacks a structure. There was this attempt yet. It is informal, for lack of structure (Verbal information, emphasis added).[11]

Why former councilor believes that "if you call" the majority of the population participates, but at the same time admits that lack of structure has not encouraged people to participate in political life? Former councilor takes into account that "we have not here a tradition, a culture of strong participation, and society always expects the government to solve everything." This interpretation relates the political attitude of the majority of the population and the system of conservative actions, persuading the people to "gigantic and modern works," which often does not benefit the people and that he participated in the so decision planning the city, old-fashioned attitude, using political coercion to enlist the society, especially the municipal civil service, to support the government's actions. This same system institutionalized welfare action as the only possible policy for municipal management, not encouraging people to participate in public policy development, transforming it into customer dispatches' policies.

According to the former councilor Paulo César, "by the time we have this planning structure that we have mechanisms, such as the day of the referendum, it will participate." Before the former councilor blamed the people for the lack of participation, as if it were an individual problem on the political attitude of each citizen, now this absence is justified by the planning framework of weakness and

[11]An interview given in April 2005.

lack of technical and political instruments to inform the people. This interpretation is superficial because it is not lack of infrastructure, especially the media, but it is the misuse that makes these means.

According to the advisory council of communication[12] (at the time), the city government had a contract with eleven providers of sound and advertising services: Roque Publicidade, RBT Pedro Martins, Walson Publicidade, Plenitude Publicidade, Ari Soares Comunicações, Publi Som POP, Rádio Pimentel, Manchete Village of Conde, LM Publicidade, Nery Publicidade, and Industrial Publicidade. The advertising was played by a stereo with input and output microphones that allowed a broadcaster to carry out the program. Each provider was composed/made up of a central house, the seat of the service, which leaves a spinning speaker system fixed on poles, in which programming is conveyed. The catchment area of each provider is proportional to the length of the wiring system and distribution of speakers on top of the posts, which led people to call this means of communication "stake radio". The "stake radio", along with the Community Radio FM and the sound of car advertising service, they published the official program of the city. For official publications, decrees, edicts and laws, it is used the newspaper "O Liberal" and "Diario do Para", published in Belém.

The press office sends weekly to the "*stake radio*" and the Community Radio FM, a newsletter, *the newsletter of the city hall* which contains festive events, welfare actions, and utility bills, all generally promoting the ruling elite. For example, in the information for the period from 07 to 13 December 2005, reads:

> Courses: The State Representative Ana Cunha brought to the council, the government school courses and in partnership with the City Department of Social Welfare and the Municipal Department of Education, held in the period from 12 to 16 December 2005, the School Palmira Gabriel the Ethics courses in Public Service Course Waiter practical, Ceremonial and Protocol, personal and professional label, Good practices in preparation and handling of food, Pedagogic Project Development Workshop Nature and Art (Prefeitura Municipal de Barcarena 2005).

The executive and legislative power of Barcarena not use the media to inform and encourage public participation in the drafting and discussion of the plans, the budget directives law and budgets, either to make public their accounts. In short, do not use the news media to promote transparency in the administration, as stated in the LRF in Chapter IX, Section I, which discusses the Transparency Fiscal Management.

However, if it is to promote traditional management, despatches policy and transform construction works in political marketing are resources to put the ruler in evidence in the pages of state-circulation magazines. In the Journal Para Mais, in its edition 42, exposed four pages of "inaugurated Works" for managing Laurival Magno Cunha (2001–2004), and even two pages on "improvements in the building and management of the City Council."

[12]An interview given in December 2005.

Therefore, the absence or low participation of the people in the city's political life, especially in business planning, cannot be explained by the absence of the media, but the use that the ruling elite makes these means, placing them in the service of their interests and creating an administrative transparency which reproduces the system of conservative actions.

In Laurival Magno Cunha management logic had administrative transparency, as his government disclosed that "is working" on all over the territory, undertaking actions of urbanization and infrastructure, housing policy, social action sector, employment generation and income, education, among others. In short, the government is transparent because inaugurates works and undertakes actions that witness where public money is being applied.

The media are used in the service of the ruling elite, the traditional family groups, in short the ruling class, that "can capture the information, direct them to a collector center, which selects, organizes and redistributes in the interests themselves" (Santos 1993, p. 127). That little dedicator core chooses the information it disseminates, why, how, and to whom, in order to build a fabulous land management, widespread in the plans and apparently solicitous administrative programs to the problems of the population and in accordance with national law. Also manufacture the conservative administrative transparency, since use of the equipment of information media, images, and entertainment to submit to public opinion the fabulous transparency of political actions.

References

Barcarena (2001a) Lei n°1889, de 30 de Dezembro de 2001. Dispõe sobre o Plano Plurianual para o período de 2002–2005. Plano Plurianual. Barcarena, Pará

Bourdieu P (1998) Contrafogos. Táticas para enfrentar a invasão neoliberal. Jorge Zahar Editor, Rio de Janeiro

Brasil (1995) Ministério da Administração federal e da Reforma do Estado. Plano Diretor da Reforma do Aparelho do Estado. Brasília, DF

Cardoso FH (2001) Reforma do Estado. In: Pereira LCB, Spink P (ed) (Orgs) Reforma do Estado e Administração Pública Gerencial. 4ª Edição. Editora FVG,Rio de Janeiro, p 15–19

Castro F.R.X de M. (Coord) (2001) Lei de Responsabilidade Fiscal. Abordagens Pontuais. Doutrina e Legislação. 1ª Edição. 2ª Tiragem. Del Rey, Belo Horizonte

Giddens A (1999) A Terceira Via. Reflexões sobre o impasse atual e o futuro da social-democracia. Editora Record, Rio de Janeiro

de Oliveira JC, Silva PF (2000) Reformas das Instituições Fiscais: Reflexões sobre o caso do Brasil. CEPAL. División de DEsarrolo Económico. Santiago de Chile. Disponível em: http://repositorio.cepal.org/bitstream/handle/11362/7543/S00030321_pt.pdf;jsessionid=67831EEF5135F5E4C72AC57A55F25F31?sequence=1. Acesso em: 01 May 2017

Pereira LCB (1997) A Reforma do Estado nos anos 90: Lógica e Mecanismos de Controle. MARE, Brasília, DF

Prefeitura Municipal de Barcarena (2005) Assessoria de Comunicação. Informativo do período de 07 a 13 de dezembro de 2005. Não Paginado, Barcarena, Pará

Santos M (1993) O Espaço do Cidadão. 2ª Edição. Nobel, São Paulo

Santos M (2005) Da Totalidade ao Lugar. Edusp, São Paulo

Chapter 4
Conclusion

Abstract The article addresses the uses of territory and the power of backwardness in Barcarena, State of Pará, since 1980, when the production engineering system for primary aluminum was implemented by the Albrás/Alunorte Company, thus dialectically reproducing modernization and backwardness. The first part of the article presents data related to the growth of municipal public finances since the great Albrás/Alunorte project was implemented, while the second focuses on the municipal public management's distribution of these resources to political actions typical of the power of backwardness. To this effect, the article analyzes the public authorities' responses to the demands of the population, that is, political patronage and traditional planning, which served as the guidelines for the Multiannual Plan for 2002–2005 and 2006–2009.

Keywords Backwardness · Modernity · Power · Public policies · Territory

After spending an afternoon visiting the city, the tourist can go home and say that met Barcarena. If spent months in the same city, his wanderings teach him a geographical look that notes changes in the landscape and territorial configuration. It is the little surprises in the context of everyday life that have the gift of revealing angles and facets of a territory. But if studying for several years, the uses of the territory, analyzing how a system of hegemonic actions has played at every moment of the modernization process and seek to understand the geographical significance of the changes, would be haunted by the vastness of the problems.

The system of conservative actions of the traditional ruling elite of Barcarena is responsible for the reproduction of unequal and divided territory. The presence of conservative political actions in Para territory systems is not inconsiderable, to the point that we can overcome the political-administrative division of Pará 144 municipalities and another division under the influence of political action, which would produce a mapping of regional despotism. A cartography of power—the global, federal, state, and municipal level—allows comparative studies on how the system of how conservative actions act as a mediator between the tensions created

J.S. Nahum, *Modernization and Political Actions in the Brazilian Amazon*,
SpringerBriefs in Latin American Studies, DOI 10.1007/978-3-319-58030-2_4

by the vectors of transformation, coming to places which are representative of the order imposed by a global reason, previously established.

The conservative actions of the traditional political system based how to govern in Barcarena, and these actions are common to Brazilian territorial formation, especially when we get up and analyze the data that show the use of public spaces in favor of private and privatístico use. Understanding this process undoubtedly involves the understanding of territorial formation of each entity of the federation and emphasizes the complementary movement of the contradictions between the dynamics of the territory also seen on the scale of the state and the market, now globalized.

The rationalizations in administrative management, as proposed by the Fiscal Responsibility Law, did not threaten the traditional political culture hitherto reigning in Barcarena territory management. Planning instruments, transparency, and popular participation adapted to favor the reproduction of conservative actions system. However, it is necessary to analyze such instruments at the state level, interpreting how the stock system enables the institutionalization of traditional political culture as the standard management of the territory, even before establishing the Fiscal Responsibility Law. It is necessary to think how the technical densities, regulatory, informational and communicational, characteristics of the current period, provide the reproduction of old patrimonial political culture.

We realize, then, that this time of completion is just a pause for a fresh start. It is the time to evaluate the road traveled, because we need to keep walking. After all, a knowledge only increases the doubts, or in the words of Wittgenstein (1969, § 160), "the child learns by believing the adult. Doubt comes after belief".

The reordering of Barcarena territory due to the implementation of primary aluminum producer engineering system Albras/Alunorte demarcates two geographical periods in the territorial dynamics of Barcarena. We take our analysis to the period after the implementation of the Albras/Alunorte, singled out by state and corporate political action who led the implementation process of this modern production engineering system; process and period which changes all the dynamics of barcarenense territory, which now have the world time, global time, governed by the rhythm of the world market, foreign to organic solidarity, historically rhythmical by local actions.

In the sphere of political action, the discourse on modernization, social responsibility, employment, and foreign exchange for the municipality, which accompanied this reordering, are only convincing means, fables to make permissive achieving the interests of the traditional elite and the purposes of Albras/Alunorte. Such purposes guided the state action—through the Para Industrial Development Company (CDI-PA), the Institute of Lands of Para (ITERPA), the Institute of Colonization and Agrarian Reform (INCRA), the Barcarena Development Company (CODEBAR) and Japanese capital, represented by Nipon Aluminium company (NALCO)—and reorganized the territory of uses in Barcarena despite the unfavorable results for the local society that saw expropriated land, forced to migrate, leaving their production area to invent another place to live.

We have seen to emerge in Barcarena, spontaneous occupation of places and planned places, both contradictory results of actions whose purpose, far from being the target of housing policies are remedies to the tensions that arise from the incisive actions of companies that see no purpose in using the territory other than those who bring their return on investment. So also become social and political functions of the municipal seat, which stops to exercise political control over the entire territory of the uses for exercising executors' function of enterprise policy.

The Village of Cabanos, area planned to serve the company's workers, was organizationally linked to the dynamics of producer enclave of aluminum and not to place society. But the privatization of Companhia Vale do Rio Doce and the competitiveness of the international aluminum market imposed on companies desvencilharem of expenses that were not the end of the company; ranks burden of maintaining the village that was transferred to the municipal government. The city of Barcarena then went on to take responsibility for the maintenance of urban infrastructure and the village of Cabanos services. The transfer of management took over territorial form and could be evaluated considering the territorial configuration that taught to the place and the current uses that did it.

The process of installation of the engineering system in the municipality increased its revenue in the last 10 years as attracted to there, other companies, which allowed the Barcarena join the select group of 6% of Pará municipalities that have their PIB [Gross Domestic Product (Produto Interno Bruto)][1] sustained predominantly in industrial activity. Ali, industrial activity participates with 78.12% in the GDP composition, services 20.91% sector, and agriculture contributes 0.97%. Despite the increase in municipal revenue with the tax transfers on Goods and Services (ICMS) and the Tax on Industrialized Products (IPI), the social demands, even missed, press the municipal budget.

The amount of Barcarena budget is inextricably dependent on the revenue and tax transfer from the implementation of the engineering system of Albras/Alunorte. The fact that a fraction of the territory controlled by these companies is offset, so to speak, by the growth of municipal revenue by increasing constitutional transfers.

The management of Laurival Magno Cunha (2001–2004) distributed these revenues among the many of the Multi-Year Plan programs (2002–2005), prepared as we have seen, of centralizing, without transparency and popular participation. Programs displaced of the real demands of barcarense society, in which the verb build is the most used, touting a transformer entrepreneurial city in a permanent construction site that are opened or in an election year or festive dates, solely for purposes and class self-interest hegemonic and politically dominant.

The works are handed out as a gift, an official act of generosity. This image is further enhanced by orders of policy and the welfare actions. The system of conservative actions reproduces the image of the mayor and alderman are the *godparents*, the *crony*, *population bosses* to whom the city dweller, the peasant, the riverside and the migrant must resort.

The local ruling elite uses the land as a resource and plays a conservative actions system whose particularistic logic plays in Barcarena clientelistic practices that

[1]Data on gross domestic product and an increase in constitutional transfers to Barcarena are available at: http://www.sepof.pa.gov.br. Access: 08 June. 2005.

vivify the vote economics and politics of electoral alliances between families; political despotism that threatens with unemployment the public official who does not obey the orders of the dominant management and patrimonialism that dissolves the boundaries between public property and private interest. On the other hand are political actions that appropriate civil society organizations by removing the political autonomy, and appropriates the local media, especially the radio poles that have the Department of Communication of the City Hall its main customer, all transforming organizations of the population and the local media in Althusser (1999) called *ideological state apparatus.*

These expressions of traditional political culture are not unique to Barcarena, but there is a structure system of conservative actions that defines the uses of that territory in accordance with the interests of a small group, an elite who take turns in municipal governments. We can say that despite the new and modern regulation of public expenditure management standards, the ruling elite plays in Barcarena the *power of the delay* analyzed by Martins (1999), it is through the conservative actions that management modernization vector system territorial, vectors that establish decentralization and democratization, are deployed so that meet particular interests.

The ruling elite creates and recreates mechanisms to maintain their economic and political privileges. This becomes clear when we analyze the persistence of conservative actions of the Fiscal Responsibility Law implementation process in the management 2001–2004 of Laurival Magno Cunha. This government suited to their interests, planning, public participation, and transparency, mechanisms established by law to ensure the participation of the people in charge of land management, but there in Barcarena, has not gone fable of a process of modernization, since the population does not even have access to information not see improved his life, not even know the possibilities of exerting a strong citizenship.

This action type—that is just a fabulous speech—has deepened the dependency of the population in relation to welfare actions and decisions of policy. Modernizing but keeping the system logic of domination; conserve system of domination logic, using the processes of modernization. In both modernize means ensuring the privileges and the achievement of the set purposes; it consists mainly of using the more efficiently territory to carry out the purposes just over the majority.

Modernity seems to mean the same underdevelopment. The Brazilian underdevelopment is part of our historical formation built by the *class struggle in each piece of territory and the influence of these on other territories.* The class struggle is at the heart of underdevelopment and is considering using the classes that make the territory that we can understand the apparent "dichotomies between the modern and the traditional" in Brazilian territorial dynamics[2] (Oliveira 2003).

[2]Oliveira (2003, p. 60) also points out that the "originality of the expansion of capitalism in Brazil is the introduction of new relations in the archaic and reproducing archaic relations in the new, a way to reconcile the global accumulation, where the introduction of new relations in the archaic releases labor force that supports industrial-urban accumulation and the reproduction of archaic relations in the new preserves the potential for accumulation releasing exclusively for the new expansion own purposes."

I maintain that in Barcarena, despite the modern changes in land use, such as the deployment of an engineering system globally integrated and the new standards for regulating the management of public affairs, a system of conservative actions reproduced without that the demands of the entire population are met. This shows that changes in land management are not processed through disassociated decrees of the local political culture. This is primarily to reinvent the meaning of public and private, as well as the relationship of civil society with these dimensions of everyday life; therefore, the existence of equity instruments, and legal enforcement mechanisms does not guarantee by itself the democratic management of the territory.

Demand to us review the modernization project exclusive and conservative that are implemented in Brazil, and has played a greater or lesser degree, in Barcarena; question your goals, also as to which interests and land uses that such modernization has legitimized. Without this whole system of conservative actions will be played every act of modernization and strengthen the power of the delay, since it will be captured by traditional interests of those who do not see another dimension of life than the status quo. This monumental task is essential to analyze the use of the territory.

To think the territory of use is a safe geographic analysis as way to understand how citizenship is constructed or relegated to the background as seen in Barcarena. Understanding the uses we make of our territory is a condition for the construction of citizenship and a new political morality. While the territory is stage product and condition of a system of conservative actions associated with a very exclusive economic dynamics, we conform as well as users, consumers more than perfect, instead of strong people in a rich country.

References

Althusser L (1999) Sobre a Reprodução. Editora Vozes, Petrópolis (RJ)

Martins JS (1999) O Poder do Atraso. Ensaios de Sociologia da História Lenta. 2ª Edição. São Paulo Editora HUCITEC

Oliveira F (2003) Crítica à Razão Dualista. O Ornitorrinco. São Paulo Boitempo Editorial

Wittgesntein L (1969) Da Certeza. Edição bilíngüe. Rio de Janeiro Edições 70

Annex—About the Sources of Verbal Information

I conducted field research in Barcarena, intensely in the first 4 months of 2005, but it began in January 2002. It embraced representatives of civil society institutions, i.e., the organized population: Community Centre and Residents Association of Pioneiro District; Laranjal Neighborhood; Industrial Neighborhood and Bairro Novo; Community Center and Residents Association of the locality of Vila do Conde, Vila Nova, Itupanema, Sao Francisco, Cafezal, and Curupere; Education Workers Union in the State of Para headquarters of Barcarena; Union of Rural Workers of Barcarena; Guardianship Council; Social Welfare and the Municipal Council of School Feeding; a former director of Barcarena Heritage and former Secretary of Barcarena Education. We would like to rely on information from other representative bodies of the organized population, such as the Union of Construction, the Metalworkers Union, the Union of Watchmen, the Council of FUNDEF, the Municipal Council of Health, and other community centers and Association Matrix Neighborhood residents of Bairro Novo II, the district of Real Estate, but the similarities and differences of everyday life not allowed because always found myself developing an activity when they were "available" and vice versa, especially the trade unions.

In this group, verbal information was given on the condition of anonymity. Thus, throughout the text strategies are used to ensure this. Effective semi-open interviews were not made on the basis of a questionnaire or on account of life, but built around the themes: influence of Albras/Alunorte in the municipality, administrative management, people's participation in development planning and administrative transparency.

The other group that gave me verbal information is composed of members or people who joined the Laurival Magno Cunha's management (2001–2004), both acting role in the executive branch or the legislature, such as former secretary of finance management Laurival Magno Cunha and current councilman Paulo Cesar; the former management Secretary of agriculture of Laurival Magno Cunha and current councilman Alirio Cesar Magno; former district agents of Murucupi and current councilman Francisco Chaves Pereira; Renato Ogawa; Councilwoman Maria delle Grazie; and Barcarena Communication Advisor. We would like to interview the ten councilors and the mayor, but the difficulty in scheduling with

J.S. Nahum, *Modernization and Political Actions in the Brazilian Amazon*,
SpringerBriefs in Latin American Studies, DOI 10.1007/978-3-319-58030-2

them showed that more would require a few years. Given the extremely short time that these public officials have granted us interviews were structured around open questions: (1) How the Multi-Year Plan is drawn up, that is, how the goals of this plan are chosen and established?; (2) How does company participates in the preparation of this plan?; (3) How the councilors relate to barcarenense society?; (4) How the councilman relates to the secretaries and the executive?

Bibliography

Almeida AWBD (2004) Solidariedades Ameaçadas e a Redefinição do Serviço Público. In: Neves RHN, Quintela RDS, Cruz SHR (eds) (Orgs) A Reinvenção do Social. Poder Local e Política de Assistência Social em Belém (1997–2004). Editora Paka-tatu, Belém, pp 39–60

Barcarena (2001b) Lei Municipal n°1950, de 30 de Julho de 2001. Dispõe sobre as diretrizes para a elaboração da Lei Orçamentária ao exercício financeiro de 2002 e dá outras providências. Barcarena, Pará

Barcarena (2002) Lei Municipal n°1959, de 28 de Junho de 2002. Dispõe sobre as diretrizes para elaboração da Lei Orçamentária de 2003 e dá outras providências. Barcarena, Pará

Barcarena (2003) Lei Municipal n°1977, de 30 de Junho de 2003. Dispõe sobre as diretrizes para a elaboração da Lei Orçamentária de 2004 e dá outras providências. Barcarena, Pará

Brasil (2002) Constituição da República Federativa. Ridel, São Paulo

Brasil (2006a) Lei n 8.913, de 12 de Julho de 1994. Dispõe sobre a municipalização da merenda escolar. Presidência da República. Subchefia para Assuntos Jurídicos. Brasília, DF, 12 Jul. 1994. Disponível em: http://www.planalto.gov.br/ccivil/LEIS/L8913.htm. Acesso em: 29 Jan 2006

Brasil (2006b) Lei n 9.394, de 20 de dezembro de 1996. Estabelece as diretrizes e bases da educação nacional. Presidência da República. Subchefia para Assuntos Jurídicos. Brasília, DF, 20 Dez 1996. Disponível em: http://www.presidencia.gov.br/CCIVIL/LEIS/L9394.htm. Acesso em: 29 Jan 2006

Brasil (2006c) Lei n 9.424, de 24 de dezembro de 1996. Dispõe sobre o Fundo de Manutenção e Desenvolvimento do Ensino Fundamental e de Valorização do Magistério, na forma prevista no art. 60, § 7°, do Ato das Disposições Constitucionais Transitórias, e dá outras providências. Presidência da República. Subchefia para Assuntos Jurídicos. Brasília, DF, 24 Dez 1996. Disponível em: http://www.presidencia.gov.br/ccivil/LEIS/L9424.htm. Acesso em 29 Jan 2006

Brasil (2006d) Lei n 8.080, de 19 de setembro de 1990. Dispõe sobre as condições sobre a promoção, proteção e recuperação da saúde, a organização e funcionamento dos serviços correspondentes e dá outras providências. Presidência da República. Subchefia para Assuntos Jurídicos. Brasília, DF, 19 Set 1990. Disponível em: http://www.presidencia.gov.br/CCIVIL/leis/L8080.htm. Acesso em 29 Jan 2006

Brasil (2006e) Lei n 8.142, de 28 de dezembro de 1990. Dispõe sobre a participação da comunidade na gestão do Sistema Único de Saúde (SUS) e sobre as transferências intergovernamentais de recursos financeiros na área da saúde e dá outras providências. Presidência da República. Subchefia para Assuntos Jurídicos. Brasília, DF, 28 Dez 1990. Disponível em: http://www.presidencia.gov.br/CCIVIL/LEIS/L8142.htm. Acesso em 29 Jan 2006

Brasil (2006f) Lei n 8.742, de 7 de dezembro de 1993. Dispõe sobre a organização da Assistência Social e dá outras providências. Presidência da República. Subchefia para Assuntos Jurídicos. Brasília, DF, 7 Dez 1993. Disponível em http://www.presidencia.gov.br/CCIVIL/leis/L8742.htm. Acesso em 29 Jan 2006

J.S. Nahum, *Modernization and Political Actions in the Brazilian Amazon*, SpringerBriefs in Latin American Studies, DOI 10.1007/978-3-319-58030-2

Fontes E (1996) O peão de trecho e o peão de casa: a identidade operária entre os trabalhadores da construção civil de Barcarena no canteiro de obras da ALBRAS/ALUNORTE. In: Magalhaes SB, Britto RDC, Costa ERD (eds) (Orgs) Energia na Amazônia. Vol 1 MPEG/UFPA, Belém

Martins JDS (1975) Capitalismo e Tradicionalismo. Pioneira, São Paulo

Martins JDS (2000) A Sociabilidade do Homem Simples. Cotidiano e História na Modernidade Anômala. HUCITEC, São Paulo

Oliveira RMD (2000) Utopia de uma Região. SER, Brasília

Pinto LF (2004) Albrás e Eletronorte assinaram maior contrato de energia do país. Belém, mai. 2004. Disponível: www.amazonia.org.br/noticia/noticia. Acesso em: 07 Jun 2004

Prefeitura Municipal de Barcarena (1999) Assessoria de Comunicação. Subsídios para um Estudo da História do Município de Barcarena. Barcarena, Pará, [s/n], p 166

Revista PM (2005) Edição 42. Editora Círios, Belém

Ribeiro RF (2003) A Idéia de Progresso em Barcarena (1980–2002) 65f. Trabalho de Conclusão do Curso de Licenciatura Plena e Bacharelado em História. Universidade Federal do Pará. Belém

The manufacturer's authorised representative in the EU is Springer Nature Customer Service Centre GmbH, Europaplatz 3, 69115 Heidelberg, Germany. If you have any concerns regarding our products, please contact ProductSafety@springernature.com

Printed and bound by CPI Group (UK) Ltd, Croydon, CR0 4YY

01/10/2026

02225882-0001